Advanced Praise

Schuette takes the complex story of alchemy and brings a chemist's sensibility to its organization. The evolution of the mission, the understanding of individual materials, and the experimental methods are each traced through the contributions made by noted historical figures. The book is a novel take for experienced alchemical scholars and an attractive entrée for those with any standard chemistry background.
 —Brian P. Coppola, Arthur F. Thurnau Professor of Chemistry, University of Michigan, Editor, *The Hexagon of Alpha Chi Sigma*

Alchemy: the mysterious precursor to chemistry that, with its cryptonyms, illustrative language, and numerous metaphors, has for many seemed more like magic than science. In "Alchemical Matters", Keith Schuette pulls back the curtain on this ancient art to reveal the true nature of the "gods" and men that led to modern chemistry. Ever the detective, Schuette meticulously uncovers the history of alchemy and the very nature of the experiments those alchemists were performing. A must-read for any student of chemistry, historian, or alchemist at heart.
 —Nicole Moon, Ph.D. Physical Chemistry, Missouri University of Science and Technology

This historical look at alchemy from a modern scientific perspective made for a highly enjoyable read. Seeing the hidden successes of some of the earliest chemistry was a welcome reminder that inspiration can come from unlikely sources.
 —Matt Senter, Lecturer in Chemical Engineering, Missouri University of Science and Technology

Alchemical Matters: Essays on the Precursor to Modern Alchemy offers a series of interrelated accounts of several fundamental theories, substances, and apparatuses that would evolve into modern physical chemistry. Through his essays, Keith Schuette illustrates the work of prominent alchemists to explain their understanding of how each of the subject elements contributed to observable physical phenomena. Each chapter stands alone, and the combined collection weaves an interesting, cohesive anthology of physical chemistry history. A compelling collection of chemistry chronicles.

—David M. Levings, Grand Historian, Alpha Chi Sigma (professional fraternity for the chemical sciences)

KEITH A. SCHUETTE

ALCHEMICAL MATTERS

ESSAYS ON THE PRECURSOR
TO MODERN CHEMISTRY

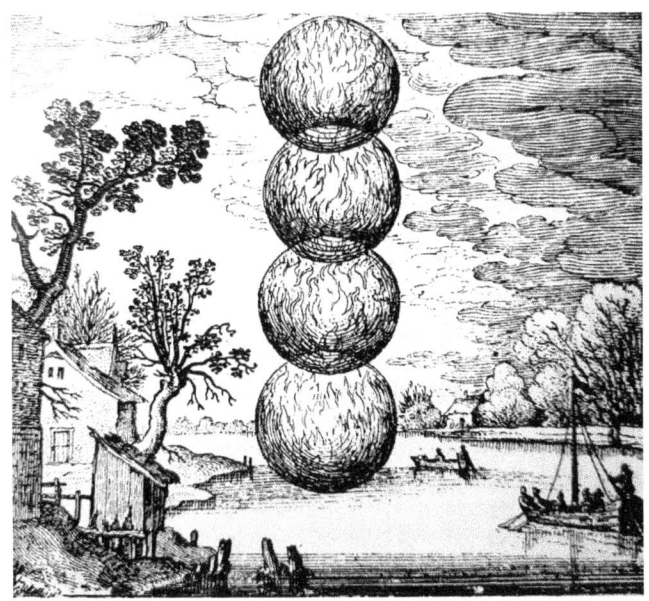

Emblema XVII *Orbita quadruplex hoc regit ignis opus* (Four Orbs govern this work of fire), Michael Maier, *Atlanta Fugiens or New Chymical Emblems of the Secrets of Nature* (Athens: Old Book Publishing, 2015), 68.

www.ten16press.com - Wauwatosa, WI

Alchemical Matters: Essays on the Precursor to Modern Chemistry
Copyrighted © 2025 by Keith A. Schuette
SC ISBN: 9781645387879
HC ISBN: 9781645387886
Alchemical Matters: Essays on the Precursor to Modern Chemistry
by Keith A. Schuette

Cover design by Dana Breunig
Edited by Dennis Uhlig

All Rights Reserved. Written permission must be secured from the publisher to use or reproduce any part of this book, except for brief quotations in critical reviews or articles.

For information, please contact:

www.ten16press.com
Wauwatosa, WI

The author has made every effort to ensure that the information within this book was accurate at the time of publication. The author does not assume and hereby disclaims any liability to any party for any loss, damage, or disruption caused by errors or omissions, whether such errors or omissions result from accident, negligence, or any other cause.

Many processes and formulas described by the author are dangerous and should never be carried out or conducted. The information presented is meant only for educational purposes.

To Sarah, her husband Dave, and their children
Elizabeth, Margaret, William, and Robert;
And to Mike, his wife Ashton, and their
children Luella and Lily

All such great blessings

Table of Contents

Introduction: Unraveling an Enigma	1
Part I: Concerning the Matter of Theories	9
Chapter 1—The Formation of the Metals: An Alchemical Perspective	11
Chapter 2—Tria Prima	20
Chapter 3—Transmutation, The Great Work	33
Part II: Concerning the Matter of Substances	67
Chapter 4—Lead, The Presence of Saturn	69
Chapter 5—Tin, Jupiter's Alchemical Influence	84
Chapter 6—Iron, Vulgar, Yet a Great Metal	117
Chapter 7—Copper, An Earthy Venus	134
Chapter 8—The Masks of Antimony	155
Chapter 9—Arsenic and Old Alchemy	166
Chapter 10—The Vitriols	206
Chapter 11—The Mineral Acids	236
Chapter 12—The Elusive Alkahest	256
Part III: Concerning the Matter of Apparatus	267
Chapter 13—Alembics	269
Chapter 14—Hessian and Bavarian Crucibles	299
Acknowledgements	306
Index	307
About the Author	315

Introduction

Unraveling an Enigma

I once worked for a man who hated talking with metallurgists. Whenever he asked a technical or operational question of me or another metallurgist, invariably, our first words would be "It depends. . . ." He wanted a simple, straightforward answer that could quickly resolve whatever issue he was confronted with. Metallurgical issues, usually, are not so simple. Likewise, interpreting or translating alchemical texts into the modern mindset is difficult. What did the alchemists mean in their writings? Well, as they say, it's complicated. There are two major issues involved. First, there is the issue of language, not in the sense of simply translating a formal language such as Latin, Greek, or Arabic—instead, the difficulty lies in the mode of expression; serious problems are likely to arise when one alchemist attempts to interpret the work of another. Secondly, in regard to working out the specifics of the chemical reactions involved, even more inaccuracies are possible as alchemists attempt to identify the actual substances in the equations and the chemical processes taking place. Suffice it to say, the obstacles encountered in reading alchemical texts are legion.

Language Problems
The alchemists used various means to express what they observed and to describe the workings behind their observations. They applied analogies, allegories, and metaphors, borrowed philosophical and/or theological terms of the day, incorporated astrological symbols and mythological images, injected color symbolism, and so on. A single alchemist may have used different terms or images for the same thing or the same term or image for different things. Metaphors might

be mixed within the same sentence. Since there was no standard technical language, as scientists use today, the alchemists adapted philosophical terms from the natural philosophers of the times or theological terms, such as Scholastic terminology, from their own religious backgrounds.

A common convention used throughout the alchemists' writings was the replacement of common terms for metals or other substances with terms from a wide range of sources. For instance, the known heavenly bodies—the Sun, Moon, Mercury, Venus, Mars, Jupiter, and Saturn—were applied to the seven known metals—gold, silver, mercury, copper, iron, tin, and lead, respectively. In place of the actual name of the metal, the corresponding planet (the Sun and Moon were included in this taxonomy), the name or symbol of the corresponding planet was employed. These names would not only be used in an objective sense–referring to the corresponding metal–but they might also be personified. Thus Venus and Mars were at times considered to be sister and brother, or lovers. Or gold was spoken of as a King whose court included the other metals.

Moreover, in regard to the heavens, operations or processes were often designated by the symbol of the corresponding astrological sign or constellation; these were twelve in number, of course. This alignment of the twelve astrological signs with twelve operations–Aries with Calcination, Taurus with Congelation, Gemini with Fixation, Cancer with Dissolution, Leo with Digestion, Virgo with Distillation, Libra with Sublimation, Scorpio with Separation, Sagittarius with Incineration, Capricorn with Fermentation, Aquarius with Multiplication, and Pisces with Projection–was common. But not all the operational taxonomies of all the alchemists listed twelve operations, so confusion was often a likely result.

In addition to drawing on the names and symbols from the heavenly realms, alchemists borrowed imagery from mythology and legends. Fantastical beasts, such as dragons, serpents, and winged lions, were often referenced. Other mythical figures–Diana, Aphrodite, Apollo, Ares, Jason of the Golden Fleece, etc.—were cited as well.

More mundane creatures also found their way into the alchemists' depiction of various substances, processes, or products of reactions. Green Lions, Black Crows, Peacocks, Eagles, Pelicans, even Wolves were incorporated into the alchemists' lexicons. And many very ordinary words were used to form analogies. For example, the Philosopher's Stone was called a medicine or a tincture. Mercury and acids were seen, simply, as water. Alembics (vessels containing the reactants and products of chemical reactions) were envisioned as eggs. An entire alchemical process might have been depicted by a tree. Often these mythological and astrological code names were woven into analogies and allegories—the use of extending a metaphor or series of metaphors throughout a story. Basil Valentine's *Twelve Keys,* for example, extended various metaphors and aliases for substances and operations to construct a narrative of his instructions for preparing "the Great and Ancient Stone." To understand his instructions, the student of alchemy was required to decipher the relationships within the imaginal language in order to unravel the required operations. By utilizing these literary devices, the clever alchemist both explained and concealed his intent. For the true adept, such language uncovered the truth, but the uninitiated or ignorant were often confounded.

In addition, much of the theoretical speculation of the alchemists incorporated the theological and/or philosophical language of antiquity and medieval times. These schools of thought included Aristotelian, Neoplatonic, and Socratic philosophies and Neoplatonic scholastic theology. Particularly, alchemists of the late Middle Ages, to avoid suspicion of heresy and deviation from the Roman Catholic Church's dogma, employed the orthodox theology of the church to explain their theories. Paracelsus, for instance, followed the opening to the Gospel of John to derive his theory of primeval matter that existed with God and was not created, but emerged from God in a similar fashion as the Divine Logos was not made but begotten. Furthermore, in his discussions of generation, he used the concept of emanationism as defined by scholastic theology, which fused the Neoplatonic concept of emanation with Christian orthodoxy.

Color and graphic symbolism, pictorial and geometrical representations, dispersions, Decknamen (aliases, secret names), paradoxes, and cryptic statements were other devices used by alchemists, which often made their observations undecipherable or confusing to their contemporaries. Colors were used not only as descriptors to distinguish substances using the same image—such as a green dragon vs a red dragon—but also to denote the stages of the transformative process that led to the Philosopher's Stone. For instance, the first stage of the process was known as the "blackness blacker than black" or the *nigredo* (Latin, "black"), followed by a phase consisting of a series of short-lived colors, named appropriately the "peacock's tail." Then the "White" would emerge, signifying the second stage—also known as the *albedo* (Latin, "white")—that produced the White Stone or White Elixir, the Stone that could change base metals into silver. The ultimate stage was the *rubedo* (Latin, "red"). After passing beyond the *albedo*, the treated matter would undergo a yellowing to become the Red Stone or Red Elixir—the Stone that could transmute metals into gold.

Dispersion (the written form of verbosity) was another technique used to confound or dissuade all but the serious adepts. Many of the alchemists' writings contained extraneous descriptions, flowery language, or irrelevant phrases that were meant to discourage all but the most dedicated from practicing the alchemical arts. Some of these pre-scientists purposely tried to hide their real intentions, thus protecting their work and, perhaps, their economic status from those who would appropriate their success for financial gain. (This may also have been the purpose for using Decknamen—aliases or codes—for substances and processes.)

The language of alchemy became even more complicated–and obscure– when certain writers depicted certain alchemical principles and operations pictorially. The diagrams of Michael Maier are good examples. One of his most widely published drawings was his "Squaring the Circle" from *Atalanta Fugiens* (below):

Another example of pictorial obfuscation, found in an illustration from the 1695 *Opera omnia of Philalethes* (below), involved the purgation of stibnite (see the chapter "Masks of Antimony").

INTRODUCTION

Purgatio materiae et reduction geniti crudi in Genitorem coctum,
ut urina sua lavet mercurium

Geometrical representations and graphic symbols were also among the non-verbal conventions employed by alchemists. Perhaps the most referenced geometrical representation used in the alchemical works was Aristotle's design of the four elements and the four qualities:

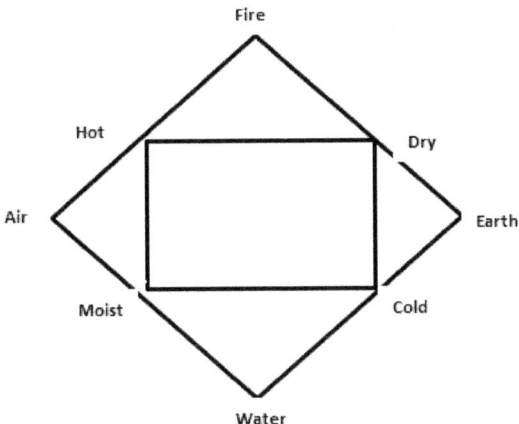

Aristotle's Elements and Qualities

Graphic symbols that designated the metals were quite commonly employed but were not always universally understood. The works of Johann Hollandus (Flemish alchemist of the 16th and 17th centuries) used this convention—and likely confused many of his readers.

> *... make an Aqua fort of* ☽, ☉, ♁ *and* ✳ *am; and to a quarter lb of this AF., add* ℥ *ij* ✳ *; dissolve in it ij lb* ☿*; draw off the Aqua fort per Balneum, and you will thus kill* ☿*.*
>
> (Johan Hollandus, *The Mineral Work, The R.A.M.S. Library of Alchemy Vol. 26*, R.A.M.S. Publishing Co., 1980, 171.)

Paradoxical, cryptic, and riddle-like descriptions provided additional hurdles. For instance, a paradoxical description for the Philosopher's Stone was "a stone that is not a stone" (attributed to Zosimos). Gold was said to be both Fire and Water (allusions to its golden-yellow color and its fusibility—associated with the flowing nature of water). One of the most famous cryptic riddles came from the words attributed to Hermes Trismegistus: *That which is above is from that which is below, and that which is below is from that which is above, working the miracles of one* (Hermes Trismegistus, *The Emerald Tablet*). So it was not enough that synonyms, aliases, and secret names (Decknamen) replaced simple, more precise terms. The alchemists manipulated their words into mysterious, enigmatic riddles as well.

Compounding these issues was the problem of interchanging terms—using different terms to denote the same substance or process. For instance, Martinus Rulandus, in his *A Lexicon of Alchemy* (pp. 220-225), lists over 100 terms for *prima materia* or over 30 designations for lead. On the other hand, the same term might be employed for different things (sometimes appearing within the same text). A good example of this befuddling practice was the use of "Mercury" and "Sulfur," which could refer to either the elements or the principles that composed substances.

Finally, the most challenging concerns I encountered in writing this book were the difficulties common to any translation from one language to another. Many original alchemical texts were already translations themselves, usually from Greek or Arabic into Latin; others were translations from one of these languages into the more vernacular German, French, English, or Italian. Since

INTRODUCTION

it is rare that the direct translation of a single word is likely to preserve both the specific meaning and all the associated connotations of that word, I often found it necessary to use several words or phrases to accurately convey the specific meaning and capture the connotations as well. For the purposes of this book, I was reliant on sources that translated the texts into English. Consequently, in an attempt to accurately convey the specific meaning and associated connotations, I made an effort, whenever possible, to compare at least two different translations, and, in the cases where they existed, I often considered several texts from the same author.

The Problem of Chemistry
But ultimately, chemistry itself–the complex study of matter, its properties, and how substances combine with or separate from one another, the rudimentary science with which the alchemists grappled–was my major problem. Just trying to accurately identify the substances and reactions the alchemists wrote about often seemed impossible. Even if a term was clearly defined by one alchemist, another alchemist might have meant something completely different by the same term. Or a single alchemist might have clearly identified a specific term in one work, and then, in another work, defined it in a completely different way. Such issues were complicated when an alchemist–often to protect his work– omitted steps or directions, leaving me to guess his intention.

Then there was the additional layer of complexity involved in bringing alchemical texts into the modern chemical universe. Direct translation of the materials mentioned by the alchemists into today's chemical equivalents (even when the material was clearly identified) was very difficult, since the alchemists' description of reaction products would not always match the results obtained using modern reagents. The presence of impurities in the materials at the disposal of the alchemists would have been a likely culprit. Even in the slightest amount, impurities could result in completely unexpected outcomes, and, more likely than not, the substances used in the alchemists' investigations were not in their purest form, being imperfectly refined from their ores or contaminated in some way. So, in addition to identifying the particular substance or compound, the possible ores or minerals from which the substance was derived also had to be identified. For example, to understand the outcome of a reaction, it was not only necessary to identify a metal—such as iron, copper, or tin—but the ore or mineral from which the metal was derived also had to be considered to account for a particular unexpected color in the reaction product.

The Intent of *Alchemical Matters*

I first encountered the enigma of alchemy during my metallurgical and chemical studies in college. More particularly, when I joined the professional fraternity for the chemical sciences, Alpha Chi Sigma, I was required to engage with the history of alchemy as the precursor of pre-modern chemistry. Like many other students at the time, I did not have a clue what the alchemical writings meant. But I was intrigued. When I learned how difficult the subject was, I set it aside as did many of my colleagues. It was just too esoteric, I thought; it would require the patience and tenacity of a Hercule Poirot. I did not have the time.

Until now.

Alchemical Matters began as an endeavor to increase my own understanding of the history and chemistry of the alchemists. My hope, however, is that this work will also be an aid to others who have an interest in the subject. As such, it has been an interesting exercise in rendering the methods and theories in the alchemists' writings amenable to the modern mind. I should also mention that several of the essays included here are dependent on the works of such historians of science and chemistry as Vladimir Karpenko, Lawrence M. Principe, Matteo Martelli, William R. Newman, and others, who have made significant contributions in bringing the works and lives of the alchemists to light. The essays also include several of my own speculations which I hope will point out possibilities, stimulate thought, and emphasize the difficulty of unraveling the enigma of alchemy.

A Note on Conventions

Names of the metals are not capitalized unless they appear as such in quoted material or as names of substances coined by an alchemist. Terms such as Mercury, Sulfur, and Salt are capitalized when referring to principles. Otherwise, they are not capitalized when referring to the elements or compounds. Also, the Aristotelian elements of Air, Water, Fire, and Earth and the qualities of Hot, Dry, Wet, and Cold are capitalized. Quotes have been lightly edited for easier reading, including capitalizations, etc.

Part I

Concerning the Matter of Theories

Chapter 1

The Formation of the Metals: An Alchemical Perspective

Introduction

Mercury and Sulfur, as a couple, would most likely strike us as a rather odious pair. Mercury, a liquid, silvery metal, we know to be quite toxic and the substance that lies behind the phrase "mad as a hatter."[1] Sulfur, a solid, yellowish chalcogen, on the other hand, does not fare much better in common parlance. It is associated with the smell of rotten eggs and the awful aroma of Yellowstone's geysers and "stink pots"—a name well deserved.[2]

Mercury   Sulfur

In spite of their notorious reputations in polite company, these two elements have very important benefits. Mercury, for instance, is used in thermometers, barometers, and button batteries. Sulfur, combined with other elements and compounds, has wide applications as well: car battery and fertilizer manufacture, oil refining, water processing, and mineral extraction. Other applications for Sulfur-based chemicals include rubber vulcanization, paper bleaching, and the making of such products as cement, detergents, and pesticides. However much we may benefit from these applications, we still tend to look down our noses (and in the case of Sulfur, hold our noses) at these members of the Metal and Chalcogen families.

But Mercury and Sulfur were not always held in such low esteem. Many centuries ago, they were highly regarded and understood much differently than we understand them today—not as two elements with their own distinctive properties, but as two principles. To the early alchemists, Mercury was related to Aristotle's moist exhalation while Sulfur corresponded to his smoky exhalation. These two "exhalations," according to Balinus,[3] and later Geber, who recapitulated Balinus's theory, emanated from the earth's center and condensed underground to form stones, minerals, and metals, including gold. According to Geber, these principles combined in different proportions and degrees of purity to form the metals.[4]

Before proceeding further, we should define certain terms. Western alchemists relied heavily on Aristotelian concepts in the development of their theories of matter—especially on two terms Aristotle used to explain how substances formed—*mixis* (mixture) and *compositio* (composition). In Aristotle's view, *mixis* meant a homogeneous combination of ingredients, and *compositio* meant a mere juxtaposition of uncombined parts. In the modern understanding of these two terms, the definitions have been flipped: a mixture, today, is understood in the mechanical sense as an adjoining of uncombined ingredients, whereas a compound is defined as a united homogeneous combination of ingredients joined by chemical bonds. Interestingly, as will be seen below, a transition was made by which the understandings of these two Aristotelian terms were exchanged, paving the way for the current interpretation.[5]

Aristotle's Elements

In addition, it should be remembered that Aristotle's physical system considered that substances were able to exist in both liquid and solid states, that they were composed of elemental water or a mixture of the elements earth and water, with heat or cold serving as the solidifying agents. Aristotle saw the fusibility of the metals as indicative of a liquid component and their solidity as evidence of their elemental earthiness. Thus, the metals were seen as composite substances, composed generally of "watery" and "earthy" material.[6]

The Geberian View
Alchemists used two theories, in conjunction, to explain how gold and other metals could be products of Mercury and Sulfur. The work of the medieval alchemist Geber (most likely the 13th-century Franciscan Paul of Taranto) was an example of how these theories were understood and applied. In his *Summa Perfectionis*[7], Mercury and Sulfur were compounds comprised of some combination of the Aristotelian elements—fire, air, water, and earth. According to this theory, the elements were minute corpuscles that bonded to form larger, more complex corpuscles, united in a "very strong composition."

> *In general, we now say that every one of these is of a most strong composition and uniform substance; and that, because the earthy parts in them are through their least particles* [per minima] *united with the airy, watery, and firy; so that in resolution no one of them can be separated, but each with all and everyone is dissolved, by reason of the strong union which they have each with other, in their least particles....*[8]

This early theory, known as the corpuscular theory of matter, set the stage for the mechanistic understanding of matter, promulgated by Robert Boyle, Isaac Newton, and others, as the composition of atoms—discrete particles—which merged to form the elements.[9]

The second theory involved the generation of the metals from Mercury and Sulfur. The *Summa*'s explanation of how Mercury and Sulfur combined to form the metals did not understand these metals to be elements in the modern sense. These corpuscular compounds were composed, instead, of more minute corpuscles of the Aristotelian elements. In addition, the description of the bond as a "very strong composition" revealed a move away from the Aristotelian *mixis* to *compositio*, treating Mercury and Sulfur as compounds.[10] The Mercury-Sulfur theory of the formation of metals regarded metals as composed of a mercuric essence which rendered them fluid, and Sulfur which gave the liquid metal stability. In this theory, the Sulfur corresponded to the dry and solid qualities of a metal, while the Mercury provided the moisture and metallic character. (John Norris—Charles University, Prague, Czech Republic; Ph.D. in History and Philosophy of Science—suggested that the Sulfur–Mercury theory may have been derived by generalizing the process that congealed cinnabar when Sulfur and Mercury were combined under appropriate conditions).[11]

In a vaporous state, the Mercury and Sulfur corpuscles—though differing in purity, size, and relative proportions—constituted a certain range of particles. Heating in a vessel would allow a volatile fraction of the particles to escape and enable the corpuscles to form compound particles. According to Norris, the *Summa* purported that the Mercury-Sulfur dyad "... balanced each other's opposing qualities as [mixing occurred], leading to a degree of qualitative equilibrium in the resulting metallic corpuscles." The metals were all, in essence, composed of Mercury combined and coagulated with Sulphur. The differences in their accidental qualities,[12] were postulated to be a result of differences in their Sulphur, caused by a variation in geologic and soil formations and in the extent of their exposure to heat.[13] Gold resulted from the perfect combination of the principles of Mercury and Sulfur in their purest forms and exact proportions. But if either the Mercury or Sulfur, or both, were impure or mixed in the wrong ratio, baser metals would result. This theory became the basis for the theory of transmutation of the metals. If the metals were constituted by only two ingredients, which differed only in their purity and proportions, then, purifying the Mercury and Sulfur and manipulating their proportions, it was believed, would produce gold.[14]

Evidence

The theory was more than speculative, however. It had an experimental basis, as well, which was founded in the process of the sublimation of Mercury (quicksilver) and Sulfur (brimstone). Two observations supported the claim for the corpuscular nature: The sublimed Mercury collected in the aludel,[15] or sublimatory vessel, as tiny droplets, and the Sulfur precipitated as finely divided powder. Thus, the process of sublimation, as interpreted by Geber, revealed the particulate structure of Mercury and Sulfur. Secondly, he observed that these two substances could be sublimed intact with little or no residue left in the bottom of the aludel. Together, these observations led Geber to understand that the Mercury and Sulfur particles remained unaltered and could not be reduced further to other compounds.[16]

Application

According to this theory, noble metals were composed of relatively refined and tightly packed particles. Gold, for example, was postulated to be made of uniformly small corpuscles (particles) of Mercury and Sulfur. This minuteness allowed the particles to be packed tightly, which rendered the interstices between the particles to be very small, as well. Since the particles of Mercury were "... very pure, they coalesced tightly based on the principle that like goes to like and does not sublime." Further, it was purported that the smallness of interstitial gaps resulted in the relative heaviness (density) of gold compared to other metals such as iron and copper (For comparison, the specific gravity of gold is 19.31, for silver—10.49, for iron—7.86, and for copper—8.92):[17]

> *Sol is created of the most subtle substance of Argentvive, and of more clear fixture; and of a small substance of sulphur clean, and of pure redness, fixed, clear, and changed from its own nature . . . That gold is of the most subtle substance of Argentvive is most evident, because Argentvive easily retains it. For Argentvive retains not any thing that is not of its own nature. And that it hath the clear and clean substance of that is manifest by its splendid and radiant brightness, manifesting itself not only in the day, but also in the night. And it hath a fixed substance void of all burning sulphureity is evident by every operation of it in fire: for it is neither diminished nor inflamed. . . . Therefore the most subtle substance of Argentvive brought to fixation, and the purity of the same, and the most subtle matter of sulphur, fixed and not burning, is the whole of gold. But in it is found a greater quantity of Argentvive than of sulphur; wherefore Argentvive hath greater ingress into it. For this cause, whatsoever bodies you would alter, alter them according to this exemplar that you may deduce them to the equality thereof. The way to effect which we have now given. For gold having subtle and fixed parts, those parts could in its creation be much condensed: and this was the cause of its great weight.*[18]

This explained the heaviness of gold and accorded with Geber's assertion that iron and copper were much more porous than gold or silver, since, he surmised, they contained heterogeneous earthy particles that disrupted the packing of the mercury corpuscles (the iron and copper referenced were most likely their ores).[19]

Gold   Iron   Copper

Experimental results supported this understanding that copper and iron were less tightly packed than gold. Exposed to an intense heat, impurities within the metals were driven off, and the iron or copper were converted to a cinder or a powder.[20] According to The *Summa*, the internal porosity allowed these metals to be penetrated by the fire of calcination. When this penetration occurred, the fire drove off the earthy sulfur contained in the metal. In contrast, gold could not be calcined by mere fire, which supported the inference that gold was composed of a tightly packed, purer mercuric essence, coalesced ("fixed") by sulfur:

> *Of Venus and Mars, the way of calcination is one; yet diverse from the former, by reason of the difficulty of their liquefaction. And it is this, either of these bodies reduced into plates, must be heated red-hot, but not melted. For, by reason of the great quantity of earthiness in them, and the large measure they have of adjustive and flying sulphureity, they are easily this way deduced into calx. And that therefore is, because by reason of much Earthiness, mixt with the substance of argentive, the due continuation of argentive is disturbed. Therefore, porosity is caused in them, through which the sulphureity passing may fly away; and the fire, by that means having access to it, burn and elevate the same. Whence it comes to pass, that the parts are made more rare, and through discontinuity of the rarity converted into ashes.*[21]

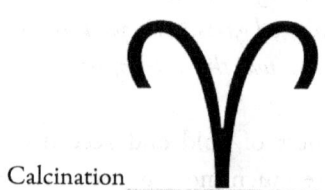

Calcination

Conclusion
Today we may be amused by the early alchemists' corpuscular theory of matter. But their inferences, derived from observations and experiments, were, nevertheless, rather insightful and not far off the mark. The basic concept within the theory—that matter was composed of discrete particles, which combined to form more complex compounds—was, in essence, the modern understanding. Alchemical thought contained the seeds from which the early modern chemists would develop their more advanced theories on the formation of matter. The alchemists' theory of the formation of matter, however, was not their only contribution. Their methods for arriving at their conclusions were just as important. Alchemical thought was not only speculative, as modern understanding of its history would have us believe. There was a strand in the alchemical tradition that believed theory should be derived from experiments and direct observations. Thus, the early alchemists also established one of the foundational pillars for the current scientific method.

Notes

1. Mercury was used in hat making to toughen the fur's fibers and make them mat together more efficiently. The compound used to moisten the fibers was Mercury Nitrate Hg $(NO_3)_2$. Hatters, breathing in the fumes of the Mercury nitrate, would suffer from Mercury poisoning, which could result in neurological and reproductive damage.
2. The term "chalcogens"—Periodic Table Group 16—was derived from the Greek word *chalcos*, meaning "ore formers." The group consists of the elements: oxygen, sulfur, selenium, tellurium, and polonium. See Chemistry Explained. Chalcogens- Chemistry Encyclopedia- structure, elements, metal, gas, number, name, salt (chemistryexplained.com).
3. A 9th-century alchemist known for his work *The Book of the Secret of Creation*.
4. Lawrence M. Principe, *The Secrets of Alchemy*, (Chicago: The University of Chicago Press, 2013), 35.
5. William R. Newman, *Atoms and Alchemy: Chymistry & the Experimental Origins of the Scientific Revolution*, (Chicago: University of Chicago Press, 2006), 26-28.
6. John Norris, "The Mineral Exhalation Theory of Metallogenesis in Pre-Modern Mineral Science." *Ambix*, March 2006.
7. This work has been attributed to the alchemist Abu Musa Jabir ibn Hayyan, more commonly known as Geber, but most likely, the actual author was the 13th-century alchemist Paul of Taranto, a Franciscan from southern Italy. How explicitly Geber purported the corpuscular nature within his version of the Mercury-Sulphur theory has been debated. Antonio Clericuzio in *Elements, Principles, and Corpuscles: A Study of Atomism and Chemistry in the Seventeenth Century* argued that Geber only vaguely referenced corpuscularism. In contrast, William R. Newman in *Atoms and Alchemy* contended that Geber was much more explicit.
8. E.J. Holmyard and Richard Russell, *The Works of Geber,* Kessinger's Legacy Reprints, (Kessinger Publishing), 56.
9. William R. Newman. See also, Antonio Clericuzio, *Elements, Principles, and Corpuscles: A Study of Atomism and Chemistry in the Seventeenth Century*, (Dordrecht, The Netherlands: Kluwer Academic Publishers, 2000), 2-3.

10. Newman, 27-28.
11. Norris.
12. An Aristotelian term which included intrinsic qualities like color or hardness, and attributes such as place, position, length, relation to other things, and actions being undertaken.
13. Norris.
14. Principe, 36.
15. A pear-shaped vessel for chemical sublimation. Aludel is the Arabic word for the Latin sublimatorium. See Mark Haeffner, *Dictionary of Alchemy*, (United Kingdom: Aeon Books, 2004), 43.
16. Newman, 29.
17. Newman, 32-33. Also, the principle "like to like" most likely had its origin in the work of Pseudo-Democritus and was also quoted in the Turbo philosophorum; see William R. Newman, *The Summa Perfectionis of Pseudo-Geber: A Critical Edition, Translation, and Study*, (Leiden: Brill, 1991), 715 n112.
18. E.J. Holmyard and Richard Russell, *The Works of Geber*, Kessinger's Legacy Reprints, (Kessinger Publishing), 130-131.
19. Newman, 33.
20. The process, known as calcination, involved exposing a substance to intense heat until it became a dry, powdery substance. Geber defined calcination as the conversion of a "thing" into a powdery substance by the removal of the humidity consolidating its particles.
21. Holmyard, 105-106. See also, Newman, 29-33.

Chapter 2

Tria Prima

Introduction

In the early 16th century, a new theory of matter entered onto the alchemical stage. Theophrastus von Hohenheim, (b. 1493) popularly known as Paracelsus (i.e., above Celsus),[1] introduced his theory of matter, which he called Tria Prima—the three 'things'—around 1515. This theory would later be given the more sophisticated name of the three principles, encompassing the principles of Mercury, Sulphur, and Salt. His innovation was an expansion of the two principles of metals (Mercury and Sulfur) theory, and yet the Tria Prima did not simply involve an addition to the earlier Arabic dyadic theory.

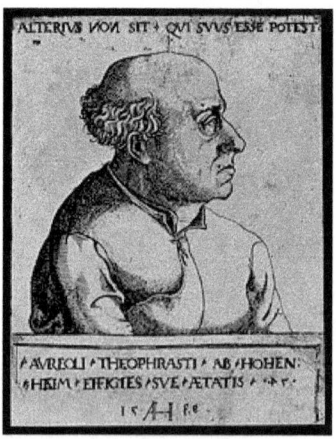

1538 portrait by Augustin Hirschvogel

CHAPTER 2

Mercury-Sulphur-Salt

The beginning of all material things, Paracelsus asserted, was not the group of four elements of Aristotle (Earth, Air, Fire, and Water) but the "three principles,"—the Tria Prima—of Sulphur, Salt, and Mercury. For Paracelsus, these were symbolic categories as much as they were rudimentary components of matter. Mercury, the volatile and metallic constituent of bodies, carried the quality of liquidity, accounted for preservation, and was believed, in reality, to transcend the liquid and solid states. Paracelsus used terms like liquor and cataronium (self-invented) to describe Mercury's characteristics.

Sulphur accounted for combustible natures; it denoted expansive force, evaporation, and dissolution. Seen as the flammable, combustible aspect of matter, it was associated with fire. Sulphur was also considered the fluid connecting the High and the Low, and thus in more spiritualized or occult writings, became connected with spirit and the Holy Spirit of the Holy Trinity in particular. Other terms Paracelsus gave to Sulphur included "ignis," "feur," "harz," and "resinam," and he attributed oleaginous qualities to the principle and accredited it with providing body, substance, and structure.

Salt represented an unburnable, nonvolatile ash or earth. It was base matter—the solid, permanent element—and represented contractive force, condensation, and crystallization.[2] In addition, Paracelsus attributed balsamic (healing) properties to salt, which he also applied to Sulphur. Other properties he assigned to Salt included alkalinity, solidity, coagulative forces, and the source of all color.

As already noted about Sulphur, Paracelsus attributed spiritual qualities—as well as material characteristics—to these three principles. According to his theories, each pointed toward a particular state of matter: Mercury towards the fluid-smoky state, Sulphur to the inflammable-fatty state, and Salt to the solid-crystalline state. In this triad, Sulphur performed the function that united two opposites; it became the soul that joined spirit (Mercury) to matter (Salt).[3]

The Tria Prima

The Origins of Tria Prima

Paracelsus's three-principle theory was much more complex than simply asserting that all matter was constituted by some combination of the principles of Mercury, Sulphur, and Salt, however. He did not start with the metallic Mercury-Sulphur theory of the early alchemists. Instead, the derivation of his theory began in the philosophical-theological concepts of Neoplatonism and Christian theology. Paracelsus envisioned the genesis of the Tria Prima in a hierarchy of creation with God at the apex, extending the Divine through several stages of creative emanations.

He rejected the traditional theories of matter used by physicians prevalent in his day and offered a new perspective on the leading theoretical concerns at the time. He was not able to eliminate every vestige of traditionalism in his rejection or achieve complete consistency in his theory, however. For instance, he utilized the Arabic Mercury-Sulphur principles accepted by the popular consensus, and he could not entirely eliminate Aristotelian and Galenic concepts.[4]

Yet, his approach was both challenging and constructive, and—relying on Neoplatonic and orthodox Christian theological concepts—it was in harmony with his theological beliefs, especially those of emanationism and Trinitarianism. Though not actually new, Neoplatonism was experiencing a renewal at the time, and the rediscovery of Neoplatonic sources injected a novel and controversial element into the philosophical discourse. The application of Neoplatonism to a theory of matter appeared innovative, credible, and appealing to the avant-garde—particularly advantageous for advancement in the intellectual and medical circles.[5]

Paracelsus's initial perspective on the theory of matter was the conventional church teaching on *creatio ex nihilo* (creation out of nothing), but his theories were infiltrated by emanationism,[6] derived from Stoic, Neoplatonic, and Gnostic sources. Two works, *On Minerals* (*Liber mineralium* publ. c. 1570) and *Labyrinthus Medicorum* (1537-38), provided an understanding of the emanationist element of his theory.

> *God created all things, something from nothing. This something is a seed; the seed contains the end of its predestination and office. And ... there is nothing that is created in its final form, but vulcan must complete it ... all things are created as prime matter and after that the vulcan follows and turns them into ultimate matter through the art of alchemy.* (Labyrinthus Medicorum).[7]

CHAPTER 2

> *The principle, then, was first of all with God, that is, the ultimate matter. He reduced this ultimate matter into primal matter. It is just the same way as the fruit, which is to produce other fruit, has a seed. The seed is in the primal matter. So, in the case of minerals, the ultimate matter is reduced to the primal, as in the case of a seed. The seed here is the element of water. God determined that there should be water. Then He conferred upon it, besides this nature, that it should produce ultimate matter, which is in the water.* (On Minerals)[8]

The starting point of his theory followed the thought of the fourth gospel in which creation was affirmed through the Divine Logos. According to Paracelsus, primeval matter—he also called it "ultimate matter"—existed with God, and was not created but emerged from God in a similar fashion as the Divine Logos was not made but begotten—an essence, a hypostasis.

Paracelsus's "primeval matter," however, had two different meanings in his writings—the Prime Matter of the World and the Prime Matter of Individual Objects. Water was the prime matter of Individual Objects, the matrix or seed of minerals and metals. The Prime Matter of the World as a whole was the Word (Paracelsus's term was "Fiat"). Not matter in the modern sense, of course, this "Fiat" was a Logos (an immaterial spirit which existed by itself, substantially, in the beginning and was reminiscent of the Prologue to the Fourth Gospel: "In the Beginning was the Word . . .") This elaboration kept his theory within the bounds of the Church's orthodoxy, both Catholic and Protestant.

Primeval matter, then, was not created, but emanated from God at the initiation of Creation, and only thereafter did the Word—Fiat—become material and apprehensible. Out of first matter, as an *Iliastrum* (Paracelsus's term that corresponded to the biblical logos), the Three Principles: Salt, Sulphur, and Mercury, emerged. The Divine Logos was not simply the uttered word, but a rational content of thought which corresponded to the ultimate reality of the universe, and connoted Law in Nature. As such, the Word existed substantially, not merely spoken, but as a hypostasis distinguishable from God and yet remaining with God.

The next stage, according to the gospel of John, was the manifestation of God's word in Life and Light, followed by the Logos becoming flesh. Similarly, Paracelsus visualized the materialization of the pure spirit as the next stage. How the word Fiat became material Paracelsus asserted, was through the emergence, in the spiritual-Iliastrum, of the Three Principles (Tria Prima)—Salt, Sulphur, and Mercury.[9]

Why Three Principles?

But how did Paracelsus decide on three basic principles? Paracelsus recognized the need to replace the elements of Aristotle with concepts that were more compatible with his hypostatic principles and more capable of playing an explanatory role.[10] How to incorporate the basic building blocks of nature using Aristotle's principles presented an obstacle, but the traditional, classical model, which embodied the Aristotelian 'four' symbolism, was a powerful unifying factor and remained too familiar and persuasive to be completely rejected. In response, Paracelsus relegated the four elements, the four Aristotelian qualities, and the four humours of Galen to limited roles, even to the point of their being lost in a free-for-all of his self-created terms such as the vulcanus, the iliaster, the aniadus, the ilech, and the archeus.[11]

Paracelsus drew on two sources to reach his new conception: Trinitarian theology and alchemy along with applied chemistry.[12] Imbued in the religious worldview of his milieu, Paracelsus was drawn to a tripartite solution. He speculated that the Creator had intentionally granted precedence to the number three. As a consequence, this number applied in the most fundamental ontological contexts and lay behind all aspects of creation. The Trinitarian principle was in the essence of the deity, in the Word of God, and therefore present throughout creation. Since the number three was ontologically fundamental to Paracelsus's belief system, he found it essential to adopt three basic principles for his theory of matter.

Paracelsus argued that his Three-Principles made perfect sense in both the microcosm and the macrocosm.[13] His Tria Prima was the product of Divine Fiat, and, as such, was as valid as Aristotle's four elements. For Paracelsus, the greatest wonder of creation was that God had separated all things into the Aristotelian four spheres of existence, each reflecting the powers of an underlying material existence, and each with a threefold character. In other words, each was created in the likeness of the Trinity. Paracelsus saw confirmation of his view in Genesis, which affirmed that human beings were created in the image of God, understood to be Trinitarian.[14]

From his knowledge of alchemy and applied chemistry, Paracelsus then identified Mercury, Sulphur, and Salt as the tripartite basis for his theory of matter, which he derived from his knowledge of alchemy and applied chemistry. This was particularly advantageous to the dissemination of his theory, since these entities were already significant in the theories of medieval alchemy. More importantly, however, Paracelsus believed that Mercury, Sulphur, and Salt provided a valid explanation for the nature of medicine at the time. Each existed in many physical forms. Each corresponded to an identifiable aspect of human identity. Salt represented the body; Mercury represented the spirit (imagination, moral judgment, and the higher mental faculties); and Sulphur represented the soul (the emotions and desires).[15]

In addition, Paracelsus regarded the Tria Prima as possessing the full range of properties needed to describe all the chemical processes that occurred in nature. According to his theory, the three principles closely approximated the chemicals of the same names in their native form, allowing salt to be a generic term. The attractiveness of the Mercury-Sulphur-Salt model was its effectiveness in explaining physical change and the ease with which each of these substances could be shown to exist in many physical forms since they all possessed an extensive range of physical properties. Moreover, these three principles readily accounted for complex phenomena—even changes of color, for example—in both organic and inanimate contexts.

Furthermore, he attached equal weight to each of the principles, which contrasted with the ideas of the alchemists, who saw them predominantly as part of a hierarchy. Even those alchemists who subscribed to the Sulphur-Mercury theory of metals regarded Mercury as the metallic principle, while Sulphur was relegated to the status of an impurity. Salt, the importance of which was established through biblical authority and through its diversified utility, was assigned an even lower place in the alchemical hierarchy. Paracelsus recognized the biblical and practical basis for the adoption of Salt as a principle, but the model for this primal substance was not confined to common salt (sodium chloride). Instead, his imagination was captured by a wide range of salts, especially nitre (potassium nitrate), and the vitriols (sulfates) of copper, iron, and magnesium, as well as the diverse and biologically active salts generated by metals such as mercury, arsenic, antimony, and lead.[16]

Paracelsus's opinions were based on more than his theoretical reasoning, of course. He also produced a great deal of experimental evidence, which he believed would prove to a reasonable person the correctness of his theory. One of his experiments, for example, involved the burning of wood. The burning process, he averred, was the work of Sulphur; the smoke was the work of Mercury; and the residual ash was evidence of Salt.[17]

Once adopted as a basic tenet, the tripartite principle was reiterated throughout Paracelsus's writings. He applied it as a major unifying theory whether he was discussing organic, inorganic, macrocosmic, or microcosmic contexts.

Pieter Van Sompel's undated engraving of Paracelsus, based on a painting by 17th-century Dutch artist Pieter Soutman (Science History Institute).[18]

Etiology of Disease

Although he was familiar with traditional alchemical ideas about the constitution of metals, Paracelsus developed his three-principle theory specifically in the context of his attempts to establish a unified approach to the etiology of disease.

As indicated in the discussion of the origin of the Tria Prima, a crucial dimension of the three-principle theory was its capacity to account for natural phenomena at every level in the cosmos. As seen from the discussion on the origin of the Tria Prima, Neoplatonic cosmology, Paracelsus believed, entailed a unified conception and continuous hierarchical gradation that emanated from the Divine realm to the realms of nature and material existence.

Essential to Paracelsus's theory of disease—as well as to his other theories—this unified conception of the cosmos assumed active interaction between the firmament and humans or other living organisms. He envisioned innumerable bonds linking the entire system of sympathy between higher and lower states of existence, which facilitated the interaction of various spirits or demons or non-personalized entities, and exercised seminal or energized functions, and served as potential sources for disease.[19]

As Above So Below—A Chymical Worldview

Paracelsus sought a "chymical worldview" in his endeavor to apply his theory to disease.[20] He saw God as the master chymist, who created an ordered world out of the primordial chaos, and further, he related the processes of creation to the chymist's extraction, purification, and elaboration of common materials into chemical products. In addition, Paracelsus held that God's final judgment of the world by fire was likened to the chymist using fire to purify metals. This meant that all natural processes were chymical processes. For example, the cycle of rain through sea, air, and land was for Paracelsus a great cosmic distillation. The formation of minerals underground, the growth of plants, the generation of life-forms, as well as the bodily functions of digestion, nutrition, respiration, and excretion were inherently chymical processes.[21]

To develop this "chymical worldview," however, Paracelsus tapped into another source for the further explication of his theory—the Hermetic tradition. In this tradition, God produced matter by separating materiality from essentiality. Hermes Trismegistus visualized the formation of the universe as a process that takes place in God: "There is not anything of all that hath been, and all that is, where God is not," he proclaimed.

Trismegistus continued:

> Hence: "The matter, son, what is it without God, that thou shouldst ascribe a proper place to it?" There is no matter that is not made active (energeitai-"actuated")-"but if it be actuated, by whom is it actuated" other than by God ... "for we have said, that Acts or Operations are the parts of God ... Whether thou speak of Matter, or Body, or Essence, know that all these are acts of God. And that the Act of Matter is materiality and of the Bodies corporality and of Essence essentiality; and this is God the whole. And in the whole there is nothing that is not God [*The Divine Pymander* of Hermes Mercurius Trismegistus].[22]

Hermes Trismegistus, from *Viridarium chymicum*, D. Stolcius von Stolcenbeerg (1624)

In this view, processes like distillation, sublimation, putrefaction, and solution could also be used to divide a naturally occurring substance into its three primordial principles of Mercury, Sulfur, and Salt. Creation of the world was a process of separation from the "great mystery"—the substance of the Divine—that produced the three principles of Sulphur, Salt, and Mercury, which further separated into the elements, from which were born all the "earthy, watery, airy, and fiery things of the world."[23]

Paracelsus believed that Sulphur, Mercury, and Salt contained the poisons contributing to all diseases.[24] Just as everything in the macrocosm was born out of the Tria Prima, diseases were also born into these categorical principles and manifested themselves corporeally as saline (outbreaks of the skin, for example), sulphurous (inflammations or fevers of various sorts), or mercurial (usually diseases associated with an excess of moisture, such as phlegm or bodily fluids generally).[25] By understanding the chemical nature of the Tria Prima, a physician might discover the means of curing the disease. With every disease, the symptoms were caused by one of the three principles.[26] Thus, for Paracelsus, each disease had a separate cure depending on how it was contracted, either by the poisoning of Sulphur, Mercury, or Salt. The treatment, then, required the identification and separation of the offending poison.

According to Paracelsus, this separation would leave behind the toxic elements of the substance. Once purified, the Tria Prima could be recombined to yield an "exalted" form of the original substance, free from impurities and toxicity, and capable of operating as a medicine. Paracelsus gave this process of separation and reintegration the name *spagyria* from the Greek words *span* and *ageirein,* meaning "to draw out" and "to bring together."[27]

Paracelsus believed that he had discovered a theory that explained both the origins of disease and the means by which illnesses could be treated.

Rejection of Tria Prima
Paracelsus's favored Tria Prima theory did not enjoy preeminence for long, however. Joan Baptist van Helmont (1580-1644) was one of the first alchemists who rejected the Paracelsian links between macrocosm and microcosm and refused to think of the Paracelsian first principles as preexistent in material substances. Sulphur, Salt, and Mercury, he argued, were instead generated in substances by the application of heat.

By the latter seventeenth century, van Helmont's acids and alkalis had come to be regarded as both chemical substances and as opposing chemical principles, which accounted for all possible chemical and physiological reactions. From this perspective, the theory of acids and alkalis became a viable alternative theory for the four elements of Aristotle and the three chemical principles associated with Paracelsus.[28]

Later, Robert Boyle (1627-1691) pointed out that there was no real agreement among Paracelsian alchemists as to which properties in mixed bodies resulted from which of the three principles. Boyle questioned whether any physical matter should be considered a principle or an element. He viewed all substances as different configurations of a particulate common matter. For Boyle, Aristotelian elements and Paracelsian principles were not adequate to explain the results of his experiments. What was needed, he proposed, was a particulate view of matter in which the smallest particles obeyed physical laws determined by God.[29] Though Boyle's views were not the final nail in the Paracelsian coffin, his theory marked at least the beginning of the funeral procession for the Paracelsians.

Conclusion

The Tria Prima resulted from Paracelsus's endeavor to create an entirely new world system, embracing the whole of theology and natural philosophy, which he offered as an alternative, and possibly a substitute, to the prevailing contemporaneous systems of medicine and natural philosophy. His theory of matter provided a terrestrial, material trinity that reflected the celestial, immaterial Trinity, and the human triune nature of body, soul, and spirit. Though his approach and assertions stirred controversy, his efforts offered a fresh and novel perspective on the nature of matter and the etiology of diseases, and captured a significant following throughout the 16th century and into the following century. Within a generation, however, his beloved Tria Prima was questioned, subjected to critical scrutiny, and–by the 17th century–rejected by the majority of the emerging chemists.

Notes

1. A 2nd-century philosopher whose philosophy was a blend of Platonism, Aristotelianism, Stoicism, and Pythagoreanism. See Thomas, Stephen (2004), "Celsus", in McGuckin, John Anthony (ed.), *The Westminster Handbook to Origen*, Louisville, Kentucky: Westminster John Knox Press, pp. 72–73.
2. Anne Marie Helmenstine, Ph.D., "The Three Primes of Alchemy," retrieved 15 October 2022 from "Three Primes of Alchemy (Paracelsus Tria Prima)," (thoughtco.com). See also, Charles Webster, *Paracelsus: Medicine, Magic, and Mission at the End of Time*, (New Haven, CT: Yale University Press, 2008), 138-139.
3. Walter Pagel, "The Prime Matter of Paracelsus," *Ambix: The Journal of the Society for the Study of Alchemy and Early Chemistry*, October 1961, 117-135.
4. Galen (126-216)—A Greek philosopher, physician, and surgeon in the Roman Empire.
5. Charles Webster, *Paracelsus: Medicine, Magic, and Mission at the End of Time*, (New Haven, CT: Yale University Press, 2008), 132.
6. The tension between creationist and emanationist positions was reconciled in medieval theology, allowing Paracelsus to employ the Neoplatonic hierarchy of hypostatic entities, which were applied with great imagination to bestow dynamic vitality to every part of the macrocosm and microcosm (See Webster, 133). Emanation, from the Latin *emanare* meaning "to flow from" or "to pour forth or out of", is the mode by which all creation comes from a first principle or God as referred to in many religions. In emanation, all things are derived through steps of degradation, at each step becoming a lesser degree of the first principle or God. (Retrieved from Wikipedia 7 Nov 2022).
7. Pagel, *The Prime Matter of Paracelsus*.
8. Paracelsus, *A Book About Minerals in The Hermetical and Alchemical Writings of Paracelsus, Vol. I*, edited by Arthur Edward Waite, (Mansfield Centre, CT: Martino Publishing, 2009), 241-242.
9. Pagel, *The Prime Matter of Paracelsus*.
10. Webster, 133.
11. Aniadus—the spiritual activity of things; Ilech—first matter, the beginning of everything; Archeus—a vital force—the immaterial principle that governed animal and vegetable life. The formative power of

nature, which divides the elements and forms them into organic parts. The principle of life. The power which contains the essence of life and character of everything; Vulcanus—the transcendent blacksmith. The fire which is the vehicle to acquire knowledge. See Dictionary.net and Selfdefinition.org.
12. The exact sources from which Paracelsus drew were not clear, according to the Paracelsian biographer Charles Webster. However, according to an accusation of Andreas Libavius (from the generation following the self-proclaimed great doctor—Libavius was born in 1540, a year before Paracelsus's death in 1541), Paracelsus's portrayal of his Tria Pima theory may not be as original as he claimed. Andreas Libavius asserted that Paracelsus plagiarized the three principles of Mercury, Sulphur, and Salt from Isaac Hollandus, a Dutch or Flemish alchemist whose identity was unclear. According to Libavius, the real source of the doctrine of the three principles, their inter-dependency, and separation into the elements, came from Hollandus's work *De opera vegetabili*. Libavius further observed that Paracelsus's gradations of medicines were almost identical to what alchemists such as Arnold, Lull, and Rupescissa enumerated. As further evidence of Paracelsus extracting others' works, Libavius noted that Paracelsus was not the first to make a tincture of antimony since Basil Valentine and Roger Bacon had previously described it. Whether Paracelsus appropriated the Mercury-Sulphur-Salt conception of the structure of matter from Hollandus or not, he worked this tripartite understanding into his overarching theory of matter. He did not simply add Salt to the Geberian theory of Mercury and Sulphur. The addition of a third member to his tripartite theory was necessary to meet the requirements of his theological and philosophical reasoning. See Bruce T. Moran, *Andreas Libavius and the Transformation of Alchemy: Separating Chemical Cultures with Polemical Fire*, (Sagamore Beach: Watson Publishing International, 2007), 155.
13. This is reminiscent of the dictum of Hermes Trismegistus in the *Emerald Tablet*—"What is above is like what is below, and what is below is like that which is above. To make the miracle of the one thing." (Twelfth Cen. Latin Translation), Dr. Jane Ma'ati Smith, C. Hyp. Msc. D., *The Emerald Tablet of Hermes and The Kybalion*.
14. Webster, 135.
15. Webster, 139-142.

16. Webster, 133-138.
17. Ibid.
18. Elizabeth Berry Drago, "Paracelsus the Alchemist Who Wed Medicine to Magic." https://www.sciencehistory.org/distillations/paracelsus-the-alchemist-who-wed-medicine-to-magic. Retrieved 15 December 2022.
19. Webster, 142-143.
20. Lawrence M. Principe, *The Secrets of Alchemy*, (Chicago, The University of Chicago Press, 2013), 128-129.
21. Principe, 129-132.
22. Walter Pagel, *The Prime Matter of Paracelsus*.
23. Bruce T. Moran, *Distilling Knowledge: Alchemy, Chemistry, and the Scientific Revolution*, (Cambridge, MA: Harvard University Press, 2005), 72-76.
24. The first mention of the Tria Prima was in Paracelsus's work, *Opus paramirum*, dating to about 1530.
25. Moran.
26. Webster, 139-142.
27. An example of the application of spagyria and the concept of Tria Prima can be found in Paracelsus's *On the Nature of Things*. There, Paracelsus asserts that plants and animals can be "resuscitated" by chemical means. The "chymist" was instructed to burn wood, mix the ashes with a distillate extracted from the same kind of wood, and then leave the mixture in a warm place until it became slimy. The "chymist" was then further instructed to allow this "slimy material" to putrefy and be buried, from where it would grow into a tree of the same type that provided the wood, "but more powerful and noble than before." This process was seen as a spagyric preparation of wood that was subjected to analysis (separation) by means of distillation, and the three ingredients were identified as the Tria Prima of the wood. See Lawrence M. Principe, *The Secrets of Alchemy*, (Chicago, The University of Chicago Press, 2013), 129-132.
28. Otto Tachenius, a student of Franciscus del le Boe Sylvius (1614-1672), was so enamored with the idea that he claimed that acid and alkali were the "architectonic instruments of nature" found in sublunary things. See Bruce T. Moran, *Distilling Knowledge: Alchemy, Chemistry, and the Scientific Revolution* (Cambridge, MA: Harvard University Press, 2005), 72-145.
29. Ibid.

Chapter 3

Transmutation, The Great Work

Introduction
In the alchemical mind, nature always moved to fulfill its perfect teleological purpose—seeds sprouted and grew into the mature plant; fertilized eggs developed and grew into the mature, adult animal; human embryos developed and matured into adult human beings. These observations led many alchemists to deduce that base metals such as lead, tin, and iron would naturally transmute into the more noble metals of silver and gold. If these processes could be properly understood, the alchemists reasoned, perhaps the natural processes that moved all animate and inanimate things to their teleological end could be reproduced—perhaps even accelerated—by art in the alchemist's laboratory. Out of this hope arose the endeavors to develop the art of transmutation of the metals.

This art was not a single pursuit, however. Each alchemist took his or her own path, but many built on their predecessors. Though these paths represented variations of the art, they had many concepts, theories, and methods in common. The commonalities and differences are explored in this discussion.

Underlying Assumptions or Primary Theories
The alchemists, natural philosophers, and artisans of metals (metallurgists of the day), shared underlying assumptions or theories about nature. According to Raphael Patai—ethnographer, historian, Orientalist, and anthropologist—in his work *The Jewish Alchemists* (1994), there were three basic theories the alchemists held in common—the unity of nature; the analogy between the growth

of plants and animals, and the development of inanimate matter in the mineral realms; and the analogy between the human body and soul and the bodies and souls of metals.[1]

Unity of All Nature
The theory of the Unity of All Nature purported that "... all the visible forms of matter, whether mineral, vegetable, animal, or human, were manifold forms of one basic essential substance...." Though they appeared different, they were variant manifestations of the one essence that existed in all animate and inanimate objects. From this theory came the notion that a base metal could be transmuted into a noble metal by introducing a minute amount of the essence of the noble metal into the base metal. Once the essence of the precious metal had been infused into the base metal, which was assumed to be in an unhealthy state (Here, we see the third theory, discussed below, at play), the sick base metal would gain its health by being transmuted into the noble healthy metal.[2]

Theory of Analogy (Inanimate Matter Likened to The Growth of Plants and Animals)
Closely related to the theory of the unity of nature was the analogy drawn between the development of plants and animals and the development of minerals and metals within the earth. From the perspective of this analogy, the vegetable and animal realms provided a window into the workings of nature. Mature plants originated from seeds; birds emerged from fertilized eggs; and the human infant emerged from the mother's womb to later become an adult. Since the alchemists viewed these natural processes as functions of their ends or goals, they believed over time the metals and their ores would transform from the lowly, base grades into the more exalted, precious ones, eventually becoming gold. Thus, the alchemists believed that if they could discover this natural process, they would be able to reproduce it in their laboratories and speed up the transformation.[3]

Another Theory of Analogy (Metals Likened to Human Bodies and Souls)
Not actually a separate theory, but more of a corollary to the previous theory, was the analogy between the human body and soul and the bodies and souls of minerals and metals. Zosimos (a Greco-Egyptian alchemist who lived at the end of the 3rd century or the beginning of the 4th century) reported that Maria the Jewess (also an alchemist of the same time period) believed that "Just as

man is composed of four elements . . . (and) . . . just as man results (from the association of) liquids, solids, and the spirit, so does copper."[4] Evidence of the application of this analogy is also seen in the classification systems, not just in the theories, of such alchemists as al-Razi (Persian physician, philosopher, and alchemist—c. 865-925 CE) who listed the minerals in six categories—spirits, metals, stones, vitriols, boraxes, and salts—and al-Khwarizmi (alchemist and polymath, c. 780-850 CE). For instance, in his *Book of Secrets,* al-Razi considered mercury a spirit while gold, silver, iron, and copper were bodies:

What one must know about Substances
Part One: Substances Required for the Chemical Art
1. *Concerning substances, there are three classes: animal, vegetable, and mineral. But the minerals fall into six groups: spirits, metals, stones, vitriols, boraxes, and salts.*
2. *Of spirits, there are four: mercury, sal ammoniac, sulfur, and arsenic sulfide.*
3. *Of metals, there are seven: gold, silver, iron, copper, tin, black lead, and Chinese iron.*
4. *Of stones, there are thirteen: marcasite, magnesia, iron ore, tutia, lapis lazuli, malachite, turquoise, hematite, white arsenic, kohl, talc, gypsum, and glass.*
5. *Of vitriols, there are five: black vitriol, white vitriol, yellow vitriol, red vitriol, and green vitriol.*
6. *Of boraxes, there are six: borax of bread, natron, borax of goldsmiths, tinkar, borax from Zarawand, and borax of willow.*
7. *Of salts are there eleven: good salt, bitter salt, mountain salt, Andarani salt, naphtha salt, Indian salt, soda, urine salt, salt of ashes, salt of lime, and salt of egg.*[5]

When it came to mercury, Al-Khwarizmi's interpretation was slightly different. Like al-Razi, he considered pure mercury to be a spirit, but quicksilver (a mineral of mercury) he labeled a body:

The bodies: They are gold, silver, iron, copper, lead, tin, and quicksilver, which is a strange substance like jelly or paste . . . The spirits: They are sulphur, arsenic, mercury, and ammonia. The bodies were so named because they remain stable when exposed to fire, but these

> *are called 'spirits' because they 'fly' [evaporate] if touched by fire . . . arsenic of several types: red, white, and green; the green is the most deadly; the best of them is the 'flaky' (Safaa'iHii). . . .*[6]

Further, in this theory, metallic bodies were conceived as animated objects. In Greek and Latin alchemical writings, the generation of metals within the earth was described in a similar fashion to the generation of living beings. For instance, the anatomy of plants was used to illustrate the characteristics of metals as in Pseudo-Democritus's *Four Books* where metals and minerals were referred to as "plants" while their colors were described as their "flowers."[7]

Additional evidence of this theory was found in the alchemical understanding of mercury as a living body. It was reasoned that if metals were alive, they must be ensouled. According to Aristotle, the soul was the source of movement of a body. In this way, a magnet was seen to have a soul since it attracted iron, making the iron move. This reasoning occurred as well in the understanding of mercury as an ensouled body. When mercury was spilled, the liquid metal formed small globules that continuously rolled on flat surfaces, appearing to move on their own. Such movement was seen as meeting Aristotle's definition of an ensouled body.[8] The demonstration that the mercury moved on its own proved to the alchemist that the metal was alive with a body and soul. This observation, coupled with the belief that living bodies and spirits could be transmuted into other bodies or souls (similar to the concept of transmigration of souls),[9] led to the belief that metals could also be transmuted.

From this theory, came the belief that bodies and spirits could be changed into one another and back again, and by harnessing this process, the alchemical adept could transmute lower metals into higher ones.[10]

Concepts

In addition to these underlying assumptions, the alchemists adopted particular concepts—such as Aristotelian Matter and Form, Atomism, *Minima Naturalia*, and *Semina Rerum*—to explain their theories of transmutation.

Matter and Form

According to Aristotle, all physical objects were compounds of matter and form, which became a doctrine known as hylomorphism (a portmanteau of the Greek words "*hule*," matter, and "*eidos*" or "*morphe*," form). Form did not mean shape as it might imply to the modern mind. To Aristotle, "form" was the essence or definition of a thing or object. Natural philosophy, he contended, was concerned with changes in matter, of which there were two kinds—accidental changes and substantial changes. To the alchemists, these types of changes were key to determining if a true (i.e., substantial) transmutation had occurred. In their writings, the alchemists pointed out that some results were transmutations of appearances only, that is, they were the result of accidental changes, while other products of transmutations were substantial. These concepts of changes were employed to distinguish those transmutations that only appeared to have changed a base metal into a noble metal from those transmutations that led to a product different from the initial ingredients. For instance, when iron was placed in a solution of copper sulfate, the iron, being more reactive than the copper, would combine with the sulfate ions to form iron sulfate, while the copper would precipitate out. The alchemists took this as evidence that iron could be transmuted into copper and concluded that a substantial change had taken place.

Accidental changes involved qualities and quantities. For instance, polishing shoes would be an accidental change since the shoe itself does not change; only the property of its color changes. Or if a person cuts his or her hair, the person does not change. Only the property of hair length does. For a substantial change to occur, according to Aristotle, three criteria had to be met: 1) an underlying essence, principle, or thing must persist, 2) a "lack," which is one of a pair of opposites, must be present, and 3) a form must be acquired during the change. In an accidental change, the underlying thing is the object or substance that acquires a new accidental property. The underlying thing, in the previous example, is the person or shoe that persists through the change. In other words, in accidental changes, the underlying object or substance remains, but the form does not change.

On the other hand, a substantial change required matter and form to account for the change. Form is the "what" that unifies matter into a particular object. In substantial changes, then, the substance or object passes away to be replaced. Here, the underlying thing that persists is the matter of the substance or object. For instance, when a person dies, the matter of the body persists but takes on a new form. (This is not a change of shape as the term "form" may, at first, imply.)[11]

Atomism

In early Greek natural philosophy, the pre-Socratics, such as Democritus, presupposed that changes in nature occurred from unchanging material principles that simply rearranged themselves, causing the observable differences in the appearances of things. These unchanging, eternal material principles were conceived as indivisible particles or "atoms" (from the Gk "*atomos*" or "*atomon*," meaning indivisible). These atoms, infinite in number, varied in size and shape, were perfectly solid with no porosity, and could not be divided further. They moved through an infinite void, colliding into one another with one of two possible results—either they repelled each other or they combined by means of minuscule hooks or barbs on their surfaces. Visible changes, in Aristotelian terms, then, were only changes of place. The atom, itself, was unchangeable, ungenerated, and indestructible. Thus, all objects, according to this conception, are simply clusters of atoms, and changes in a particular object, including growth, are a result of rearrangements of, or additions to, the atoms composing the object.[12]

Minima Naturalia

The concept of *minima* was based on the Aristotelian concept of form and the doctrine that substantial forms did not exist beyond an *ad maximum* or an *ad minimum* limit. This applied to compound bodies as well as the four elements of Fire, Air, Water, and Earth. Originally, the concept of *minima* was understood to apply to a substance that was limited as to the extent it could be divided. Below the *minima*, forms could not exist. However, the concept was interpreted in various ways by the alchemists. Medieval philosophers were divided on whether *minima* were actual parts or not. For instance, Albert of Saxony (German philosopher and mathematician, 1320-1390) interpreted *minima* in physical terms but argued that the notion of a minimum size was not absolute. He pointed out that the minimum size depended on the environment: a quantity of a substance that is too small to exist under one set of environmental conditions might be stable under another set of conditions.[13]

Geber (Paul of Taranto), in *Summa Perfectionis*, claimed there was a lower limit of size beyond which a substance or thing could not maintain its identity. It is worth noting that this did not mean he was in complete agreement with the notion of atomism. In his theory, the smallest particles were not indivisible or permanent. Rather, following Aristotle's notion of "parts" and "pores" (see Aristotle's fourth book, *Meteors*), Geber believed that substances, even solid objects, consisted of "parts and pores," too.[14] His "parts," (*pars*) denoted minute corpuscles (*minima*). Accordingly, since gold was most resistant to corruption, Geber reasoned that the precious metal contained corpuscles so small they did not permit much interstitial space,[15] which prevented contamination or penetration of other substances.

Another perspective on *minima* was offered by Daniel Sennert (German physician, 1572-1637), who conceived *minima* as atoms. His "atoms," however, were different from the classical atoms conceived by the pre-Socratics. For Sennert, atoms were particles of matter with different qualities that remained unchanged in compound bodies. He included the four elements (*homogenea*) in this theory.[16]

An additional modification to the concept of *minima* was made by Julius Caesar Scaliger (Italian scholar and physician, 1484-1558), who advanced the notion that *minima* of various substances differed in size, which, he felt, would offer a better explanation for their aggregation. For instance, according to Scaliger, the *minima* of the element Earth were the largest, while those of Water, Air, and Fire followed in decreasing order of size. This, he purported, explained the differences in states or conditions of a substance. Rain, snow, and hail, for example, consisted of Water, yet they presented in different forms. Scaliger maintained that this was due to the arrangements and connections of the *minima*. In addition, this would have explained the porosity observed in bodies since the different sizes would permit the necessary spaces. Further, the differences in size would also account for chemical reactions: the *minima* of Fire, for instance, were contained in the spaces between the other *minima* of the substances and would escape from those spaces on heating or mechanical operation, allowing the substances to combine.[17]

Semina Rerum

Semina rerum (Latin, meaning "the seeds of things") was the notion that *semina* were invisible living entities that provided a formative power or potentiality for growth. They were understood to be non-material, semi-divine entities, or as living particles of matter. In other words, they were generative forces that were responsible for the development of natural bodies.[18] As cautioned by Lawrence Principe (Drew Professor of the Humanities in the Department of the History of Science and Technology and the Department of Chemistry, and the Director of the Singleton Center for the Study of Premodern Europe at Johns Hopkins University), this understanding was not to be confused with an organic or living substance in the modern sense of those phrases. *Semina,* Principe argued, should be taken more in the sense of "... a powerful agent, an organizing principle, that works at the microscopic level to transform substances."[19]

According to the Stoics, the *semina* carried and specified the divine creative power immanent in nature that directed the rational design for all natural things. Thus, the *semina* were understood as the links between the intelligible, metaphysical world and the material, physical world, and thus, explained natural processes. For instance, Giles of Rome (Medieval philosopher and Scholastic theologian, 1243-1316) applied the concept of *semina rerum* to his theories of embryology, or Albertus Magnus (German Dominican friar, scientist, and philosopher, c. 1200 to 1280) combined the concept of *semina rerum* with Aristotelian notions regarding the formation of minerals to formulate his theory of the efficient cause (the Aristotelian agent that causes change and drives transient motion) of minerals.[20]

The concept of *semina* was prominent in many alchemical texts and was applied in various ways. The Philosopher's Stone was often seen to be the union of masculine and feminine principles—that is, masculine and feminine *semina*. It was thought that the pure seeds of gold and silver could be extracted in the form of Sulfur—the masculine seed—and Mercury—the feminine seed. These could then be injected into the base metal to be transmuted, and by virtue of their generative power transform the base metal into the precious metals of either gold or silver.[21]

The concept of *semina rerum* was central, as well, in the works of Paracelsus. Since he had rejected the Aristotelian concepts of elements and qualities, he turned to the notion of the *semina* as the invisible spiritual forces and archetypes of subsequent substances. In Paracelsus' theology, the *semina* originated in the Word and were contained in the primal matter, which he called the Yliaster, and were prior to chemical principles and elements.[22]

Still another variation of the *semina rerum* can be found in the works of Joan Baptiste van Helmont (1580-1644). He added the idea of fermentation to the concept of *semina rerum* to explain the generative power of the *semina* and the efficient cause of all things. According to van Helmont, ferments were the "parents" of transmutation, and were "The medial tools by which the semina dispose materials. . . ." In other words, the ferments are the means by which the *semina* act. Further, he maintained that the *semina* transmuted elemental Water, which he believed to be the prime matter, into the various substances of the physical world.

According to van Helmont, the ferments, aided by the division of substances into their *minima*, drove the *semina* into their respective activities. Thus, transmutation of two juxtaposed substances, reduced to their *minima*,[23] could be accomplished because their respective *semina* were free to act on one another and would result in a true mixing—or, in Aristotelian terms, a substantial change.[24] Van Helmont also incorporated the notion of masculine and feminine principles, uniting the medieval scholastic concept of *minima naturalia* with the hylozoism[25] of his teacher, Paracelsus, offering a vitalistic corpuscular theory of matter.[26]

The Instrument of Transmutation—The Philosopher's Stone
Armed with the above assumptions and concepts, along with their individual observations of natural processes, the alchemists were confident that a means for transmutation could be achieved. All they had to do was discover the precise mechanism by which the atoms or the *minima* could be recombined, or by which the appropriate *semina* were released to impregnate the desired substance for transmutation. Once the mechanism and processes were known, they could be reproduced by the arts of alchemy. This meant, however, that some agent or instrument would be required to effect transmutation. This instrument was variously known as a "medicine," the "Elixir," the "Tincture," the "it" of Hermes Trismegistus's (legendary Hellenistic period figure) *Emerald Tablet*, or as the *donum dei* (gift of God). More popularly, however, this agent of transmutation was known as the Philosopher's Stone, often abbreviated to "The Stone."

Many experiments, employing various substances and operations, were attempted to achieve the penultimate goal of finding the Philosopher's Stone and, consequently, achieving the production of pure gold. Some alchemists claimed success while others, in the end, admitted the quest was illusory. Over the course of hundreds of years, many labored to find the secret of transmutation, producing all manner of recipes by which they claimed success. Despite the variety of their recipes, most—if not all of them—followed one typical method.

Emblema XXI (Make the man and woman a circle, of that a quadrangle, of this a triangle, of the same a circle and you will have the Stone of the Philosophers)[27]

Making the Philosopher's Stone

In his book, *The Secrets of Alchemy*, Lawrence Principe (Professor of the Humanities in the Department of the History of Science and Technology and the Department of Chemistry at Johns Hopkins University) outlined what appeared to be a typical method for preparing "The Stone" before the 18th century. First, the prepared substance or mixture was placed in an alembic or retort,[28] often described as an oval glass body with a long neck (referred to as the *ovum philosophicum*). This vessel would serve as the "womb" from which the Stone would be born. Next, the neck of the alembic was sealed, usually with philosopher's clay [29] or by melting the sides of the neck together, a procedure known as the seal of Hermes, to prevent the loss of any volatile materials. The vessel with the contained substance, "egg," was placed in a furnace or otherwise exposed to heat. After a lengthy period of time—commonly specified as about a month[30]—the substance would turn black, the first of the primary colors of the Philosopher's Stone. This first stage of the operation or opus was known as the "head of the crow" (*caput corvi*), or the "blackness blacker than black" (*nigredo nigrius nigro*), or simply as the *nigredo*.

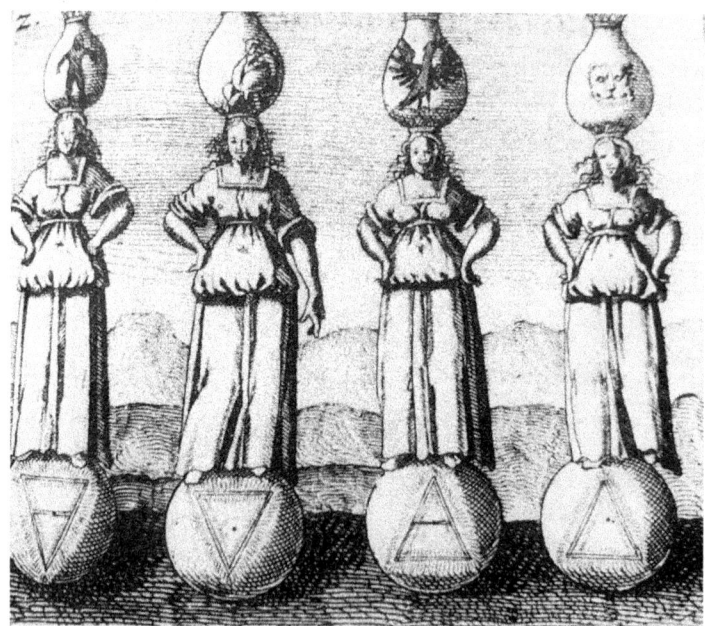

The four stages of the alchemical process. The four elements are indicated on the balls— Mylius, Philosophia reformata (1622)[31]

If the heat was continually increased over the course of several weeks, the blackness was replaced by a variety of short-lived colors known as the "peacock's tail" (*cauda pavonis*). The material in the vessel eventually became white, marking the second stage of the work, which was known as the whitening, the albification, or the *albedo*, and signified the completion of the White Philosopher's Stone or the White Elixir. The White Philosopher's Stone was used for the transmutation of base metals into silver.

Emblema XXIX (As the Salamander lives in fire, so also the Stone.)[32]

The ultimate goal was to make the Red Philosopher's Stone or Red Elixir. This was achieved by continually heating the "philosophical egg" until the white matter turned yellow and then darkened to a deep red. Once the process was completed, the retort would be broken open, and the stone removed. One final step was required, however. The Red Philosopher's Stone had to be fermented with gold in order to transmute base metals into gold. This was accomplished by "inceration"—that is, the Red Stone had to be made fusible as wax by the addition of the Philosopher's Mercury (a liquid principle), enabling the Red Stone to penetrate metals and convert them into gold.

Incercation[33]

The transmutation was accomplished, first, by placing the base metal in a crucible and melting it. Then, a small amount of the Red or White Philosopher's Stone was added to the molten metal—a process known as "projection." Finally, the contents were re-melted, achieving the final product of gold or silver, depending on whether the White Stone or Red Stone was used.[34]

A Specific Recipe

Though Principe identified a typical pattern in the preparation of the Philosopher's Stone, the method he outlined began after the primary ingredient had been prepared. These initial preparations were crucial to understanding the workings of the Stone. For example, Geber's recipes[35] offer detailed preparatory instructions.

Geber provided recipes for "medicines of three different orders." Each of these medicines, when projected on base metals, would produce corresponding "orders" of perfection. The first-order medicine produced only impermanent effects. The second medicine would produce change, but only perfect one quality, leaving the others imperfect. The third-order medicine (presumably, corresponding to the Philosopher's Stone) would perfect all qualities at once.[36] What is important to keep in mind is that Geber did not conceive of any essential difference in these medicines; they differed only in particle size. The medicines were, in fact, purified mercury,[37] with corpuscles of different sizes. To obtain the orders of the medicines, a process of repeated sublimation was required to "... attain a state of continually increasing subtlety"—that is, decreasing particle size (here "subtlety" refers to smallness). Thus, for Geber, purified, subtle mercury was the instrument of transmutation.

Sublimation[38]

First, the impure mercury (quicksilver) had to be prepared through reiterative sublimation and cleansing:

> *Of the Sublimation of Mercury and Argentvive*
> Now We will determine the whole Intention of Sublimation of Argentvive. This Work is completed, when its *Terrestreity*[39] is highly purified, and its *Aquosity*[40] wholly removed. For We are excused from the labor of removing its *Adustion*,[41] because it hath none. Therefore We say, that the Ingenuity of separating its superfluous Earth, is to mix it with Things wherewith it hath not affinity, and often to reiterate the Sublimation of it from them. Of this kind, is Talk [talc], and the Calx of Egg-shells, and of White Marble. Likewise also Glass most subtly beaten [pulverized into minute particles], and every kind of Salt prepared. For by these it is cleansed, but by other Things, having affinity with it (unless they be Bodies of Perfection) it is rather corrupted; because all such Things have a Sulphureity, which, ascending with it in Sublimation, corrupt it. And this you find true by Experience, because, when you sublime it from Tin or Lead, you find (after Sublimation) infected with Blackness. Therefore its Sublimation is better made by those Things, which agree not with it; but it would be better, by Things, with which it doth agree, if they had not Sulphureity. Wherefore, this Sublimation is better made from Calx than from all other Things; because that agrees little with it, and hath not Sulphureity.
> But the way of removing its superfluous Aquosity is, that when it is mixed with Calces, from which it is to be sublimed, it be well ground and commixed with them by *Imbibition*[42] until nothing of it appears, and afterward the Wateriness of Imbibition removed by a most gentle heat of Fire; which receding, the Aquosity of Argentvive recedes with it. Yet the Fire must be so very gentle, as that by it the whole Substance of Argentvive ascend not. Therefore from the manifold Reiteration of Imbibition, with Contrition and gentle *Assation*,[43] its greater Aquosity is abolished; the residue of which is removed by repeating the Sublimation often. And when you see it most White, excelling Snow in its Whiteness, and to adhere (as it were dead) to the Sides of the Vessel; then again reiterate its Sublimation, without the Feces; because part of it adheres fixed with the Feces, and

can never by any kind of Ingenuity be separated from them. Or afterward, fix part of it; as shall expressly be taught you in the following: And when you have fixed it, then reiterate Sublimation of the Part remaining, that it may be likewise fixed.

Being fixed, reserve it; but first prove it by fire. If it flows well, then you have administered sufficient sublimation, but if not, add to it some small part of argentivive sublimed, and reiterate the sublimation, till your end be answered; for if it hath a lucid and most white color, and be porous, then you have well sublimed it, if not, not. Therefore in the preparation of it made by sublimation be not negligent; because as it mundication (or cleansing) shall be, such will be its perfection, in projection of it upon any of the imperfect bodies....[44]

Of the mundification [45, 46] *of argentivive*
Therefore, 'tis now necessary completely to declare the Mundification of Argentvive. In order to which, we say, Argentvive is cleansed two ways; either by Sublimation, of which we shewed the way already; or by a Lavament, of which the way is this: Pour Argentvive into an Earthen or Stone Dish, and upon it pour as much Vinegar as is sufficient to cover it. Then set it over a gentle Fire, and let it heat so far, as you may well hold your fingers in it, and not more. This being done, stir about with Fingers, until it be divided into most small particles, in the similitude of Powder; and continue stirring it, until all the Vinegar you put in be wholly consumed. Then wash the Earthiness remaining with Vinegar, and cast that away: repeating the same washing so often, as until the Earthiness of the Mercury be changed into a most perfect Caelistine Color (blue), which is a sign it is perfectly washed. From these, we must now pass to Medicines.[47]

According to these instructions, the preparation involved two major steps: Sublimation and Cleansing (Mundification). Quicksilver had to be purged of its impure qualities of Adustion (dryness), Terrestrety (earthiness), and Aquosity (wetness). Since the quicksilver has no "dryness," there was no need to address that issue. Through repeated sublimation, the superfluous earthiness was removed by subliming the quicksilver (impure mercury) with substances that would not combine with the mercury but would combine with the impurities. This was followed by the removal of the mercury's wetness or wateriness (Aquosity) using a gentle heat so the mercury would not volatize; again, this was a repeated operation. According to Geber, another method might have been used for cleansing the mercury— washing with vinegar. The purpose of these reiterative sublimation and cleansing processes was to ensure that the mercury was as pure and subtle as possible. The purer and smaller the corpuscles of the mercury, the more effective they would be in penetrating the metal to be transmuted, and positioning themselves within the metal's matrix. Once particles of "The Medicine" were established, they would prevent the base metal from being volatized, and impart the desired new qualities that would effect the transmutation.[48]

(The above is from Geber's *Summa Perfectionis,* which goes on to discuss, in a speculative or theoretical fashion,[49] the preparation of the Lunar and Solar medicine of the third order for the transmutation of base metals into silver and gold, respectively.)

His work, *Of Furnaces*, provided more explicit, practical instructions for the production of the Third Order medicine and the application of the Stone: (In this work, he connected the term "the Stone" to the Third Order Medicine.)

> *Of the Way of Perfecting, according to the Third Order*
> *Having above sufficiently treated of all the ways of perfecting Imperfect Bodies, in the Second Order, we must now pass on to the Bounds of the Third Order. But what the Medicines are, and of what kind, both of the second and third Order, is plainly enough demonstrated in Our Book of the Perfect Magistry [Summa Perfectionis]; where we have with a competent and true Demonstration, shewed, that our Stone is procreated of the Substance of Argentvive: and this we did sufficiently, as in a speculative Theoretical Book. Therefore we intend here manifestly to unlock the Closure of Art, and it is thus: You must study to resolve Luna, or Sol, into its own Dry Water, which the Vulgar call Mercury: and this is so, as a Duodenary Proportion (of the Solutive Water) may contain only a part of the perfect Body. For if with gentle Fire, you well govern these, you will find (in the space of forty Days) that Body converted into mere Water. And the sign of its perfect Dissolution is Blackness appearing on its Superficies.*
>
> *But if you endeavor to perfect both Works, the White, and the Red, dissolved each of the Ferments by itself, and keep it. This is Our Argentvive extracted from Argentvive, which we intend for Ferment. But the Paste to be fermented, we extract, in the usual manner, from imperfect*

Bodies. And of this we give you a general Rule; which is, that the White Paste is extracted from Jupiter and Saturn; but the Red from Venus and Saturn. Yet every Body must be dissolved by itself in the Ferment.[50]

From here, Geber described the regimens for making the various "Pastes"— The regimen of Jupiter and Saturn, the regimen of Venus and Saturn, the regimen of Mars, the regimen of Luna, and the regimen of Mercury—key to the projection of the Stone. Since he held that the purified, subtle Mercury was the true transformative agent, its preparation was the most important:

Of the Regimen of Mercury
The Regimen of Mercury is completed in two ways. First, you must amalgamate it, well washed and purified, in the certain proportion by us underwritten. In the second way, you must distill it, and thence make an Aquavitae. For the first way, the proportion is this: Of Mercury 48 ounces, of Sol 1 ounce; of Luna 1 ounce, of Venus 1 ounce, and of Saturn 1 ounce. Melt these bodies; first the Venus and Luna; secondly, the Sol, thirdly, Saturn. Take all out of the fire, having melted them in a large crucible, and your Mercury in readiness made hot in another; and when the said metals begin to harden, pour in the Mercury, leisurely, stirring the mixture with a stick, setting it again on the fire, and taking it off, until they be all amalgamated with the whole Mercury. This amalgam, put to be dissolved for seven days, extract the water with a cloth, make the residue volatile, administering fire of ignition. This again imbibe with its whole water, and put it to be generated; and again to be dried for forty days, and you will find a stone; which put to be fixed, and you will have a stone augmentable to infinity....[51]

Multiplication[52]

Next, the sublimed and cleansed mercury had to be prepared in such a way that it could be "projected" (i.e., throwing or casting a ferment or tincture onto a substance to effect a transformation of that substance).[53] Furthermore, the mercury had to be prepared in such a way that it was "... augmentable to Infinity...," that is, it had to be multipliable.

Geber called the composition of the third order medicine "the Ferment of Ferment upon Mercury" which would then be projected onto the desired substance for transmutation:

> *Of Ferment of Ferment upon Mercury, as well for the White, as for the Red*
> *The Composition of Our Medicine, which is called Ferment of Ferment upon Mercury, is made for the White, after this manner: Take the Ferment of Luna, which is its Oyl, and add to it twice so much of Arsnick sublimed and dissolved in Water; then to both these add of Mercury dissolved, as much as the Arsnick. Mix the Waters, and set them over a Fire for one Day to be incorporated. Afterward, extract the Water by Alembeck, and revert it; this do fifteen times, so incerting,*[54] *and it will be fluid, as fusible Wax. Then add to it as much Virgins Wax melted, commix them, and project the Mixture upon Mercury washed, according as shall seem expedient to you. For that resolved is augmented in virtue and weight.*
>
> *But if this Ferment of Ferment be made for the Red: Dissolve Sol in its own Water (all the Compositions of those Waters, and of other Things, are sufficiently treated of in Our Book, Of the Invention of Perfection; wherefore We have here omitted them) to one part of that Gold dissolved, add two parts of Sulphur dissolved in the same Water together with it, and three parts of Mercury dissolved. Let all these be truly dissolved into the most clear Waters, which being mixed hot for one Day, that they may be fermented; then extract the Water fifteen times, each time reverting it. Incerate with yellow Virgins Wax; that is, with half its weight of Oyl of Blood, or Oyl of Egg: then project upon crude Mercury, according as shall seem expedient to you.*
>
> *Here note, that if you perfect this Medicine, according to the Method We have taught (in the Third Order of Our Sum of Perfection) of the Congelative Medicine of Mercury, you will find by Reiteration of the Work, and by Subtillation thereof, that one Part tingeth infinite Parts of Mercury into most high Sol, more noble than any natural Gold.*[55]

Congelation[56]

For the Stone or Medicine to work, some mechanism by which the Stone could be made to adhere or to penetrate the metal to be transmuted was needed. Thus, the Medicine was made fluid and fusible by ceration. In the instructions above, this was achieved by "incerating" the Third Order Medicine with "Virgin's Wax." This would prevent volatilization of either the Mercury or the base metal, but still allow the necessary penetration.

Specific Theories of Transmutation
To explain the observed transformations wrought by their recipes, the alchemists resorted to one or more theories— the theory of purgation by fire, the theory of the excessive Aristotelian form of gold, the theory of an intense degree of Aristotelian qualities, the theory of plusquamperfection, the theory of the existence of gold or silver *semina* within the Philosopher's Stone, and/or the theory of inversion.

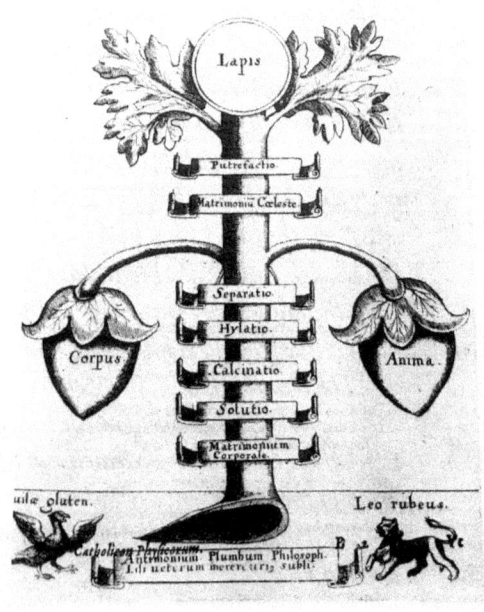

Arbor philosophica: the tree as a symbol of the stages in the transformation process—Samuel Norton, *Catholicon physicorum*[57]

Purgation by Fire (Refining)

This theory asserted that the Philosopher's Stone acted like an intense fire, purging the base metal of its impurities and superfluities that prevented it from attaining the purity of the precious metals. A common practice employed in antiquity and medieval times illustrates purgation by fire. This process was called the "dry method" (i.e., solvents were not involved) by metallurgists and was actually a cementation procedure. In the case of gold, the cementation was referred to as "cement royal" that purified a moderately pure ore. The ore was first ground and heated, forcing the silver and copper within the ore (gold ores consisted of alloys with other metals such as copper and silver) to react with a paste ("the cement"). Since gold was comparatively unreactive to the other substances present, the silver and copper would react with those substances—that is, corrode—leaving pure gold.[58]

This theory is described in al-Iraqi's (Abu Al-Kasim Muhammad Ibn Ahmad Al-Iraqi, Muslim alchemist, d. 1260) *Book of Knowledge Acquired Concerning the Cultivation of Gold:*

> ... *Silver may be converted into gold by the refinement of the smelting fire only. For it is tinctured by the fire and strengthened and transmuted and behaves like gold with the touchstone. Thus it is possible to effect a certain transmutation since the specific nature is constant.*[59]

Aristotelian "Form" of Gold

Another approach believed that the Stone was endowed with the Aristotelian form of gold. This form was projected onto a base metal, destroying and replacing the base metal's form with the form of gold.[60] Albertus Magnus, in his *Book of Minerals,* following the work of Avicenna (physician and philosopher, 980-1037), discussed this need to first destroy the base metal's form:

> *But Avicenna in his [Letter to Hasen on] Alchemy says that he found [trivial] the counterarguments of those who, in their alchemical [books], denied the transmutation of metals. And for this reason he himself adds that 'specific forms are not transmuted, unless perhaps they are first reduced to prime matter (materia prima)—the [indeterminate] matter of [all] metals—and then, with the help of art, developed into the specific form of the metal they want. But then we must say that skillful alchemists proceed as skillful physicians do: for*

skillful physicians, by means of cleansing remedies, clear out the corrupt or easily corruptible matter that is preventing good health—for good health is the end which the physician has in mind—and then, by strengthening nature, they aid the power of nature, directing it so as to bring about natural health. For thus undoubtedly health will be produced by nature, as the efficient cause; and also by art, as the means and instrument. And we shall say that skillful alchemists proceed in entirely the same way in transmuting metals. For first, they cleanse thoroughly the material of quicksilver and sulphur, which, as we shall see, are present in metals. And when it is clean, they strengthen the elemental and celestial powers in the material, according to the proportions of the mixture in the metal that they intend to produce. And then nature itself performs the work, and not art, except as the instrument, aiding and hastening the process, as we have said. And so they appear to produce and make real Gold and real Silver.[61]

Albertus maintained that the base metal must first be reduced (i.e., the form must be destroyed) to the prime matter (an indeterminate state from which all metals are derived), after which a new specific form (in this case, the form of gold) could be inhered into the resultant prime matter. The new form of gold would then replace the form of the base metal. [62]

The Role of Aristotelian Qualities

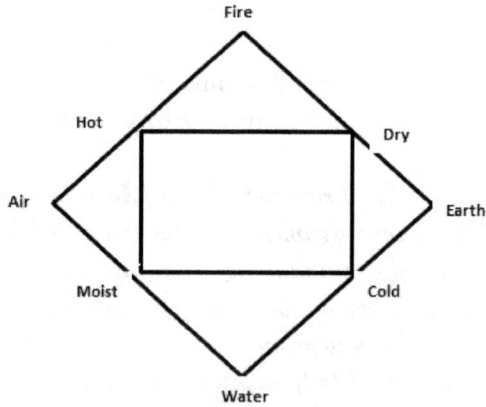

Aristotelian Elements and Qualities

In this understanding, the Philosopher's Stone possessed Aristotelian qualities such as hot and dry, which were the qualities of gold, to an intense degree. With such a preponderance of hot and dry, a small amount could overwhelm and correct the qualities of the common metal, such as the coldness and wetness of lead.[63] The *Rosarium Philosophorum* provides an example, attributed to Tudianus (his actual identity could not be determined), of the application of this notion:

> *Know that our coppery and volatile stone is in his manifest cold and moist, and in his secret warm and dry. And that coldness and moistness which is in manifest, is a watery fume corrupting and making black, destroying itself and all things, and flying from the fire. And the heat and dryness, which is in secret, is warm and dry gold, and it is a most pure oil penetrative in bodies and not fugitive, because the heat and dryness of Alchemy tingeth and nothing else. Cause therefore the coldness and watery moisture, which is in manifest, to be like unto the heat and dryness, which is in secret, that they may agree together and be conjoined. . . .*[64]

Al-Iraqi also explained transmutation as the effect of the Aristotelian qualities present in the Elixir or Philosopher's Stone:

> *. . . it has been established that the quantitative composition of these six metallic substances is constant; understand this, therefore, and know that the cause of the existence of Gold is nothing but the equilibrium of the hotness, and that is the reason why the rest of the six substances fall short of being Gold is excess either of hotness or of coldness.*[65]

According to al-Iraqi, the metallic substances consisted of some combination of the Aristotelian qualities of hot and cold. For instance, gold held these in perfect equilibrium. Silver, on the other hand, was imperfect because it consisted of excess "coldness." The other metals had relative degrees of hot and cold—copper was hotter and drier than silver, but colder than gold; though silver was colder than gold, it was hotter than lead.

> *Now to whatever of these (metallic) forms is cold, is added the hot Elixir, and it heats it and tinctures it red; while to those which are*

hot with a heat in excess of equilibrium is added the white Elixir, and it cools them and tinctures them white, and gives equilibrium to their constitution which was disordered.

And since it is known that the Elixir of Whiteness is hotter than the two Leads, in the same way that the hotness of silver is greater than that of the two Leads, then the Elixir of Whiteness may be projected upon the two Leads and will increase them in hotness and cohesion until it transforms them into the just proportion of silver and its hotness, which falls short of Gold and goes beyond the two Leads. Thus the Elixir of Silver is not excessively cold, and the Elixir of Gold is not excessively hot.[66]

Thus, to transmute silver into gold, the Elixir that was hot and red, was required to overcome the coldness. Similarly, since lead has a greater degree of coldness than silver, the White Elixir, which is hot compared to lead, will decrease the coldness of lead transmuting it into silver.

Projection[67]

Plusquamperfection

In the sphere of minerals and metals, alchemists considered the Philosopher's Stone more than perfect. It stood above perfection. It was gold (in the case of the Red Stone) or silver (in the case of the White Stone) exalted above the typical rank assigned to those precious metals. In other words, the Philosopher's Stone was "plusquamperfect." Under this theory, when the Stone was mixed with the imperfect metal, the imperfection of the base metal and the "perfect above perfect" Stone averaged out to result in perfection—that is, gold or silver, depending on the desired outcome.[68]

An example where this was at least implied can be seen in the last paragraph of "Of Ferment of Ferment Upon Mercury" from Geber's work *On Furnaces*, in which he concludes that the "Medicine" was "most high Sol, more noble than any natural Gold":

Here note, that if you perfect this Medicine, according to the Method We have taught (in the Third Order of Our Sum of Perfection) of the Congelative Medicine of Mercury, you will find by Reiteration of the Work, and by Subtilitation thereof, that one Part tingeth infinite Parts of Mercury into most high Sol, more noble than any natural Gold. (See reference note 55.)

Gold or Silver Semina within the Stone
Many alchemists also drew on the theory of *semina rerum*, and postulated that the Philosopher's Stone contained a "seed" of gold or silver (again, depending on if it was the Red Stone or the White) that was capable of transforming base metals into gold or silver.[69] The generative form of the *semina* of the gold or silver within the Stone would drive a substantial change in the properly prepared base metal, replacing the form of the impure, unhealthy metal with the form of the healthy precious metal.

Inversion
Yet another understanding found in the early Greek alchemical corpus was the idea of inversion. The Greek word often used to describe transmutations was *ekstrophe*, which literally meant a "turning" or an "inversion," and could be applied to various transformations including a coloring, a surface transformation, or a complete transmutation. *Ekstrophe*, however, was also understood to mean a process akin to the extraction of that which was hidden in a substance—that is, "a turning inside-out."[70] This understanding of transmutation is seen in a commentary (in dialogue form) to the *Four Books* of Pseudo-Democritus, referred to in the *Corpus alchemicum*[71] and entitled *The Philosopher Synesius*[72] *to Dioscorus: Notes on Democritus' Book*:

Dioscorus: How can I become intelligent, my philosopher? I want to learn the method from you; for if I try to follow what has been said, I will not have any benefit from that.
--Listen to him speaking, O Dioscorus, and sharpen your mind, Dioscorus. Look at how he says: "Turn their nature inside out: for nature is hidden inside."
--O Synesius, what transformation [lit. turning inside out] is he speaking about?
--He speaks about the transformations of bodies.

--And how can I turn it [i.e. the nature] inside out? How can I lead the nature outside?
--Sharpen your mind, O Dioscorus, and turn your attention to how he speaks: "Therefore if you perform the right treatments, you lead the nature outside. . . ."[73]

Through inversion, then, the true nature or essence of a body (metal) was brought out.

Types of Transmutations

As summarized by Vladimir Karpenko (Prof. in the Department of Physical and Macromolecular Chemistry at Charles University, Prague), the alchemists practiced several methods of transmutation that involved one or more of the following processes: Extraction or isolation of metals from a mixture of compounds, alloying precious metals with other metals, cementation, tinging base metals with precious metals, and tinging objects with a golden or silvery color. These methods were known to the craftsmen of the time and picked up by the alchemists who hoped that the method could be improved to produce a true transmutation. To comprehend the practices of the alchemists to the fullest, it is necessary to understand the transformation methods utilized by the craftsmen and tradesmen of the time. Often the only significant difference between the procedures of the alchemists and those used in the trades was that the former would complicate their procedures using their own language and ideas.[74]

Isolation of Precious Metals from Alloys

The oldest type of transmutation was that of extracting metals from their ores, or from a mixture of compounds—alloys. Extracting metals originated from the process of cupellation, which provided one of the two observations from which a theory of transmutation was developed. During cupellation, the desired metal, whether precious or base, was created by the application of heat, which gave rise to the belief that the generation of heat could accelerate the "ripening" or "healing" of metals from an immature or unhealthy state into the mature or healthy form of the precious metal. The second observation, which supported the first observation, came from miners who observed that different kinds of ores occurred together. This gave rise to the belief that minerals matured or ripened in the "womb of the earth." A common example of this was the belief that bismuth, because of its silvery-white appearance, was the immature state of silver.[75] For instance, miners gave the name *tectum argenti* (Latin, "silver being made")[76] to bismuth. This way of thinking is illustrated in the work of alchemist al-Iraqi: "thus Lead may be converted into Silver, for if you place a pound of Lead in the fire, it rectifies it and matures it, and most of it is burnt away, leaving a small part as silver."[77]

Alloying Precious Metals with Common Metals
Another practice known since antiquity, and routine among goldsmiths, was that of reducing the gold content of an alloy yet not so much as to lose its yellow color. This debasement technique could then be used to provide an apparent increase in the amount of the gold. Many alchemical recipes claimed to detail the production of gold or silver, but in actuality they were recipes for using the precious metal as an alloying agent or for producing an alloy from several ingredients that appeared as gold or silver.[78] An example of this method appeared in the *Mappa Clavicula*,[79] a medieval Latin text that contained recipes for the manufacture of metals, glass, dyes, and other craft materials:

> *Recipe #3 Mappa Clavicula*
> *Take only a little for experimentation when you do it once until you learn it thoroughly. Take 1 oz. of reddish Cyprian copper in solution [?, posios], 1 oz. of quite good silver, and melt it with chaff, until when hammered out it does not make a noise, and then melt it together with 1 oz. of gold and the same amount of natron. Then turn face to face two little bowls, i.e., two hollow earthenware pots, and put inside them the widened (i.e., hammered out) melt that has been prepared, and mix in antisma. What had been a little bead of copper is now turned into a little bead of 4 oz. of silver. In the bead we find more than an equal amount of gold. [Take] one part of Pontic sinopia, 2 parts of common salt, grind them all together, lay the sheets on the bottom and sprinkle [the sinopia and salt mixture] over them and coat them with pot clay so that they cannot breathe. Put fire under them until you feel it is all right. Take them out and you will have the very best gold.*[80]

This recipe described a process producing a golden surface to an alloy containing approximately equal parts of copper, silver, and gold. The metal would acquire a silvery-white coating as a result of annealing in air and pickling, but the final cementation in sinopia (a dark reddish-brown natural earth pigment, whose reddish color comes from hematite, an ore of iron) and salt would eventually remove everything but gold from the surface.[81]

Cementation

In cementation, alternating sheets of metals, or layers of metals and salts, were pressed together and heated, but not to such a high temperature as to melt the metals. Cementation was not a true transmutation, as some alchemists pointed out. In reality, it produced a composite material. Provided the outer layer was gold or silver, the result was passed off as a transmutation since it would appear that the initial precious metal had increased with the transmutation of the base metal to which it had been fused.[82] An example of cementation comes from the *Mappa Clavicula*, Recipe No. 8:

> *A recipe for gold—You hammer out several sheets of pure silver and lay them down onto a preparation, which will be revealed, and sprinkle on them [more of the same preparation] and melt them until they are all reduced into one. Now, this preparation is what is called Offa: Take 4 scruples of gold, 1 oz. of Macedonian glue, 1 oz. of live sulphur, 1 oz. of natron, an entire pig's gall, 2 oz. of soot, 1 oz. of Spanish minium, a whole fox gall, 1 oz. of elidrium,[83] 1 oz. of Lycian saffron, and you make a potion containing iron in which you put all this together, and with this preparation [do] as above, and you apply it underneath the sheets and sprinkle it on top. Now into 1 pound of silver you put 1 oz. of the preparation alum melt, and it will be go. . . .*[84]

Surface Treatment of Common Metals with Precious Ones

Yet another method that was claimed to be a transmutation involved imparting a golden or silvery appearance to objects made from common metals or alloys. The technique (also known as tinging, using tinctures or even the Philosopher's Stone itself) had been known since ancient times. In this treatment, the alchemist—or the craftsman—applied the secret substance to the surface and then heated the treated object—clearly an amalgamating process. For instance, gold was amalgamated with mercury, and the amalgam applied to the surface of the object. After heating and polishing, the surface acquired a golden color. Many alchemists believed that the surface treatment of an object was a substantial change or a transmutation of the entire volume of the treated material.[85] An example of this method, commonly called gilding, is found in the *Mappa Clavicula*, Recipe #55.

CHAPTER 3

The gilding of everything that you want to gild, whether it be a vessel of silver or copper:

Take an extremely thin gold sheet, shear it into tiny pieces, and put it in a mortar. Add a little quicksilver and leave it for a short time. Afterwards, add some natron and vinegar; rub it thoroughly with a pumice stone until it has the consistency of glue on account of the abundance of quicksilver. And now you put it in a clean cloth and squeeze it, so that most of the quicksilver comes out. Then you take the vessel [that is to be gilded], polish it with fine pumice, heat it, and while it cools, coat it with the amalgam, and you heat the vessel a second time and again coat it and put it on the fire. And the gold alone becomes enriched. After a little while when the color pleases you, heat the vessel and put it into a blacking liquid, i.e., the boot dressing, with which leather things are blackened. And then rub it. But if you were gilding a copper vessel, after you have polished it, coat it with liquid alum, for it will [then] receive the amalgam.[86]

Symbols of Amalgamation[87]

Methods Leading to Products of a Golden or Silvery Color (containing no precious metals)

The final method, outlined by Vladimir Karpenko, involved producing an object containing no precious metal with only the appearance of a golden (yellow) or silvery (white) color. Gold imitations, typically, were based on copper-bearing alloys,[88] while imitations of silver were based, perhaps, on tin or bismuth-bearing alloys. Another example involved using copper as the base metal that was given a silvery color using melted tin or white arsenic:

Recipe #5 Mappa Clavicula
Take copper that has been hammered out when hot, and grind filings of it in water with 2 parts of crude orpiment so that it becomes as viscous as glue, and roast it in a small pot for 6 hours and it will turn black. Take it out and wash it off, then add an equal portion of salt

and grind them together. Then roast it in the pot and watch what happens: If it is to be white, mix in silver; if yellow, mix in gold in equal portions, and it will cause wonder. . . .[89]

Transmutation Between Base Metals

The belief that base metals could be transmuted into gold and silver also led to the supposition that transmutations could take place between the base metals themselves. The most famous example of this was the exchange reaction between metallic iron and copper sulfate. (Other examples included iron into lead, copper into lead, and lead into zinc.)[90] The replacement reaction of iron and copper sulfate was referenced by Paracelsus, along with other alchemists, in his *Tincture of the Philosophers,* "Concerning the Transmutation of the Metals by the Perfection of Medicine":

> *. . . in the mountain commonly called Kuttenber, they obtain a lixivium out of marcasites, in which iron is forthwith turned into Venus of a high grade. . . .*[91]

Venus was a reference to copper but most likely consisted of chalcopyrite, a source of copper-iron sulfide.

Conclusion

The notion of transmutation throughout antiquity and the Middle Ages was a complexity of assumptions, methods, procedures, and theories. Though transmutation theories are looked upon by the modern mind as flights of fancy or as reasonable—yet erroneous—interpretations of observations, the notions served as interesting explanations of observable facts at that time and stimulated much investigation into the nature of matter. In addition, this work led to the formulation of theories and creative interpretations of established concepts, which served as a foundation for further investigations, eventually culminating in the theories of pre-modern chemistry.

Work Completed—Finis[92]

Notes

1. Raphael Patai, *The Jewish Alchemists, A History and Source Book*, (Princeton, NJ: Princeton University Press, 1994), 4-5.
2. Ibid.
3. Ibid, 5.
4. Ibid, 66.
5. Gail Marlow Taylor, *The Alchemy of Al-Razi, A Translation of the "Book of Secrets,"* (North Charleston, South Carolina: Createspace Independent Publishing Platform, 2014), 101-103.
6. Karin C. Ryding," Islamic Alchemy According to Al-Khwarizmi," *Ambix*, Vol. 41, Part 3, November 1994.
7. Vincenzo Carlotta & Matteo Martelli, "Metals as Living Bodies. Founts of Mercury, Amalgams, and Chrysocolla," *Ambix*, 70:1, 7-30.
8. Ibid.
9. The passage of a soul after death; metempsychosis, *Webster's Unabridged Dictionary of the English Language,* (New York: Random House Press, 2001).
10. Patai, 5.
11. Form vs. Matter (Stanford Encyclopedia of Philosophy). Retrieved 30-Nov-2023.
12. Berryman, Sylvia, "Democritus", The Stanford Encyclopedia of Philosophy (Spring 2023 Edition), Edward N. Zalta & Uri Nodelman (eds.), URL = <https://plato.stanford.edu/archives/spr2023/entries/democritus/>. Retrieved 21-Nov-2023.
13. Antonio Clericuzo, *Elements, Principles, and Corpuscles: A Study of Atomism and Chemistry in the Seventeenth Century, International Archives of the History of Ideas*, (Dordrecht, The Netherlands: Kluwer Academic Publishers, 2000), 10.
14. Lawrence M. Principe, *The Secrets of Alchemy*, (Chicago: University of Chicago Press, 2013), 56-57.
15. William R. Newman, *The Summa Perfectionis of Pseudo-Geber, A Critical Edition, Translation, and Study*, (Leiden: Brill, 1991), 144-145.
16. Clericuzo.
17. Ibid., 11-13.
18. Ibid.
19. Principe, 126.
20. Clericuzo, 13-16.

21. Ibid.
22. Ibid., 18.
23. Van Helmont understood minima as the smallest corpuscles capable of retaining the nature of a substance, yet beyond which, their seminal properties would still be retained. See Newman, *Gehennical Fire*, 143-146.
24. William R. Newman, *Gehennical Fire: The Lives of George Starkey, an American Alchemist in the Scientific Revolution*, (Chicago: University of Chicago Press, 1994), 143-146.
25. Hylozoism is the philosophical doctrine according to which all matter is alive or animated.
26. Newman, *Gehennical Fire*, 143-146.
27. Michael Maier, *Atlanta Furgiens, Emblemata Nova de Secretis Naturae Chymica*, 1618. Reprint Series (Athens & Manchester: Old Book Publishing, Ltd., 2015), 86.
28. See also the essay "Alembics" in this work.
29. Also known as "the clay of wisdom" and consisting of two-thirds clay free of stones and one-third a mixture of dried dung and chopped hair. See "Arsenic and Old Alchemy" of this work.
30. Often this was difficult to achieve because of the pressures that would build up in a sealed vessel, causing an explosion.
31. C.G. Jung, *Psychology and Alchemy*, translated by R.F.C. Hull, Bollingen Series, *The Collected Works of C.G. Jung, Vol. 12*, (Princeton: Princeton University Press, 8th printing of the revised edition, 1993), 229.
32. Maier, 118.
33. Philip Wheeler, *Alchemical Symbols, Fourth Edition, R.A.M.S. Library of Alchemy Vol. 21*, (Kansas City, Mo.: R.A.M.S. Publishing Co., 2018), 14.
34. Principe, 123-126.
35. Another example of detailed preparatory instructions can be found in Eirenaeus Philalethes' *Introitus apertus ad occlusum regis palatium*, Amsterdam, 1667, and a few years later issued in an English edition "*Secrets revealed: or, An Open Entrance To The Shut-Palace Of The King*," London, 1669. See the Alchemy Website, Levity.com.
36. Newman, *Summa Perfectionis*, 162.
37. Geber held to a "mercury-alone" theory in which mercury by itself was the medicine, and that repeated purification would allow the different

orders of medicine. "If you can perfect with quicksilver alone, you will be an investigator of the most precious perfection, and of that which overcomes the work of nature." See *Summa Perfectionis*, 731.

38. Wheeler.
39. Earthiness.
40. The quality or state of being moist or wet. Merriam-Webster Dictionary, Merriam-Webster.com. Retrieved 10-Jan-2024.
41. The act of drying by heat. See "adjust" in *Webster's Unabridged Dictionary of the English Language*, (New York: Random House, 2001).
42. The absorption of a liquid.
43. The reduction of a substance to a dry ash by roasting. The Alchemy Website, Levity.com. Retrieved 10-Jan-2024.
44. E. J. Holmyard and Richard Russell, *The Works of Geber, Kessinger Legacy Reprints*, (Kessinger Publishing), 86-88.
45. From the Latin *mundificatio*, meaning the act of cleansing.
46. The act of washing thoroughly. Merriam-Webster Dictionary, Merriam-Webster.com. Retrieved 10-Jan-2024.
47. Holmyard and Russell, 157-158.
48. Newman, Summa Perfectionis, 167.
49. For a complete discussion of the theory and workings of Geber's "medicines," see Newman, *Summa Perfectionis*.
50. Holmyard and Russell, 250-251.
51. Ibid., 254.
52. Wheeler, 14.
53. Projection. The Alchemy Website, Levity.com. Retrieved 12-Jan-2024.
54. Softening a substance to appear like wax, accomplished by continually adding a liquid and heating. The Alchemy Website, Levity.com. Retrieved 12-Jan-2024.
55. E. J. Holmyard and Richard Russell, *The Works of Geber, Kessinger Legacy Reprints*, (Kessinger Publishing), 255-256.
56. Wheeler, 14.
57. C.G. Jung, 407.
58. Andrew Sparling (2020), "Paracelsus, a Transmutational Alchemist," *Ambix*, 67:1, 62-87.
59. Abu Al-Kasim Muhammad Ibn Ahmad Al-Iraqi, *Book of Knowledge Acquired Concerning the Cultivation of Gold*, Translated by E. J. Holmyard, (Eastford, CT: Martino Fine Books, 2018), 12.

60. Principe, 126-127.
61. Albertus Magnus, *Book of Minerals*, translated by Dorothy Wyckoff, Prof. of Geology, Bryn Mawr College, (Oxford: Clarendon Press, 1967), 177-178.
62. Newman, *The Summa Perfectionis*, 18.
63. Principe, 126-127.
64. *Rosarium Philosophorum* (a transcription of the 18th-century English translation of the *Rosarium* in MS Ferguson 210. The text was originally printed as part II of *De Alchemia Opuscula complura veterum philosophorum...*, Frankfurt, 1550). Archive.org. Retrieved 02-Jan-2024.
65. Abu Al-Kasim Muhammad Ibn Ahmad Al-Iraqi, *Book of Knowledge Acquired Concerning the Cultivation of Gold*, Translated by E. J. Holmyard, (Eastford, CT: Martino Fine Books, 2018), 14.
66. Ibid, 16-17.
67. Wheeler.
68. Principe, 126.
69. Ibid, 126-127.
70. Olivier Dufault (2015), "Transmutation Theory in the Greek Alchemical Corpus," *Ambix*, 62:3, 215-244. See also, "Olympiodorus," Stanford Encyclopedia of Philosophy, https://plato.stanford.edu/entries/olympiodorus/#OlyAlc. Retrieved 06-Jan-2023.
71. A summary of Greek works and discoveries compiled by a Byzantine scribe in the early 11th century. It contains extracts from 24 books dating from the second to the eighth century. The oldest of these is *Physika kai mystika* by pseudo-Democritus.
72. The identity of Synesius is uncertain. The *Corpus alchemicum* does not identify him, only naming him in a list. Scholars have postulated that his identity is that of Synesius of Cyrene or Synesius of Philadelphia, but both are problematic. For a complete discussion of the possible identity of Synesius, see Matteo Martelli, *The Four Books of Pseudo-Democritus*.
73. Matteo Martelli, *The Four Books of Pseudo-Democritus, Sources of Alchemy and Chemistry*, *Ambix Vol. 60, Supplement 1*, (London: Rutledge, 2013), S129.
74. Vladimír Karpenko (1992), "The Chemistry and Metallurgy of Transmutation," *Ambix*, 39:2, 47-62.
75. Ibid.

76. Mary Elvira Weeks, "Glossary of Archaic Chemical Terms," https://web.lemoyne.edu/~giunta/archems.html. Retrieved 01 Jan 2023.
77. Karpenko, 50.
78. Ibid., 51-54.
79. The core material was originally compiled around 600 CE in Alexandria, Egypt, to which additional recipes were added over the medieval centuries.
80. Cyril Stanley Smith, Massachusetts Institute of Technology, and John G. Hawthorne, University of Chicago, "Mappae Clavicula: A Little Key to the World of Medieval Techniques," *Transactions of the American Philosophical Society, Vol. 64, Part 4,* (Philadelphia: The American Philosophical Society, 1974), 29.
81. Ibid., 30n15.
82. Karpenko.
83. A preparation of gold or silver, which is produced from two parts silver, one part gold, and one part copper. The term could also refer to Mastic (a tree) gum. See Martinus Rulandus, *A Lexicon of Alchemy or Alchemical Dictionary, Containing a Full and Plain Explanation of All Obscure Words, Hermetic Subjects, and Arcane Phrases of Paracelsus*, published in the Free Republic of Frankfurt: by Zechariah Palthenus, Bookseller, 1612, (Kessinger's Legacy Reprints).
84. Smith and Hawthorne, *Mappa Clavicula*, Recipe 8.
85. Karpenko, 55-57.
86. Smith and Hawthorne, *Mappa Clavicula*, Recipe 55.
87. Wheeler, 18.
88. Ibid.
89. Cyril Stanley Smith, 30.
90. Karpenko, 59.
91. Paracelsus, *The Hermetic and Alchemical Writings of Aureolus Philippus Theophrastus Bombast, of Hohenheim, Called Paracelsus the Great*, Edited by Arthur Edward Waite, (Mansfield Centre, CT: Martino Publishing, 2009), 28.
92. Wheeler, 85.

Part II

Concerning the Matter of Substances

Chapter 4

Lead, The Presence of Saturn

Saturn

Introduction

The concepts of affinity and correspondence stood at the heart of the worldview of the alchemists of antiquity and the Middle Ages. These concepts postulated a relationship between two different objects, which were believed to share characteristics or traits. One of the most venerable and long-lasting of these relationships was that between the seven known planets and the seven known metals of those ages (though the Moon—known as Luna—was not a planet, it was included in the list, and though other metals such as bismuth and cobalt were known to miners of those times, the seven metals familiar to and dominant in the wider culture were gold, silver, mercury, copper, iron, tin, and lead). The associations between the planets and the metals were: Sun (Sol)—gold; Moon (Luna)—silver; Mercury—mercury; Venus—copper; Mars—iron; Jupiter—tin; and Saturn—lead. Though the Saturn-Lead duo was not the shiny bauble in this august list of planet-metal correspondences, the pairing occupied a prominent role as attested to by various alchemical writings.

Lead was known to the ancients at least since 7000 BC.[1] Grey in color, it was the heaviest of the known metals, so it was often associated with slowness. Since Saturn took the longest to return to its position in the heavens, it was considered to be the slowest of the planets, and hence earned the correspondence to "heaviness" or "weightiness." Saturn was also associated with melancholia/depression, which was seen as a disorder that "weighed" on a person. This "heavy" disorder corresponded naturally to the heaviness of the metal.

Another concept of Saturn, taken from mythology, provided a further–and perhaps more significant– understanding of the nature of lead. The Roman god, Saturn, was identified with Cronus, the Greek Titan, and their myths were often co-mingled. In one story, Cronus, the old god, devoured his children because of a prophecy that predicted they would usurp his kingdom.[2] In this telling, the outward manifestation–Cronus himself–was seen as corrupt and base, while the inner manifestation–the consumed children–were seen as a potential that could be redeemed. In the alchemical mind, this notion was transferred to lead, which was considered the basest of the seven metals. Outwardly, lead was seen as a corrupt, impure material, but inwardly it contained a redeemable–incorruptible–potential. This view gave birth to the belief that lead could be transmuted into a purer or nobler metal.

Cronus (from Wikipedia)

Because lead and Saturn were so closely linked, they were used interchangeably in alchemical writings. To the modern mind, Saturn might be seen as a code for lead, but to the alchemists, the appearance of Saturn in their writings would have been understood as a direct reference to lead. In short, Saturn served as an alias for lead.

Aliases, Code Words, and Descriptions
Saturn was perhaps the most common and well-known alias for lead, but the dense metal, its ores, and other associated minerals enjoyed a variety of other code names and descriptions. Known as Plumbum (Latin),[3] lead was listed in Martinus Rulandus's *A Lexicon of Alchemy*,[4] (purported to be a "... full and plain explanation of all obscure words, Hermetic subjects, and arcane phrases of Paracelsus")[5] in association with more than 50 such code words. Some of the more obscure ones included: Scape-goat, the Dual Chibor, Draiccium, Elerator, Araxat, Alusa, Ruba, Alech, Allonoch, Alabri, Alokot, Armic, Amioch, Amitich, Araxat, Azoro, Balamba, Cartistilium, Koal, Molybdos, Mosquet dei, Molibra, Mosider, Rasas, Rasasa, Rolos, Roe, and Rocli. According to Rulandus, these were technical terms, which in themselves had no meaning, but were used only to signify lead. Equivalents from the Hermetic tradition included Aabam, Abartamen, Accib, and Aitrazat.

More clearly, lead was described by Rulandus as a "... Livid, Terrene,[6] Heavy, Metallic Body, with very little Whiteness and much of Earthy Nature," which could be converted by cleansing[7] to the more perfect metal, tin. Since lead was perceived to be the basest metal, this description reinforced the idea that lead contained an internal nature that could be transmuted to tin. According to Rulandus, lead was an "impure body containing more fixed Sulphur than Jupiter (tin)," and was "procreated from the copulation of sulphur with "Imperfect Living Silver" (mercury)."[8] This "Living Silver" was further described as "... impure, unfixed, Terrene, feculent, somewhat white on the exterior, but red inside, with a similar quality of Sulphur." Lead, then, was seen as the product of sulphur and "Living Silver:"

> *... Living Silver, which is of bad quality, gross, of bad taste, fetid, and of feeble power, like a menstruous mother, unites with a livid or leprous Sulphur, and a frigid Saturn for a son is the result, and this is Lead.*[9]

Though Rulandus's lexicon was extensive, it did not encompass all expressions or understandings of lead, its ores, and associated substances.

The Saturn in the Writings of Certain Alchemists

(Though there was agreement among alchemists in their references to lead or Saturn, differences in the terms used, descriptions applied, and understandings of the nature of lead, existed as well, as attested in the examples that follow.)

Pseudo-Geber

Pseudo-Geber (Paul of Taranto, early 14th century), in *Summa Perfectionis, Chapter XLII, Of the Alchymie of Saturn,* asserted that lead, in its own nature, could not be "... approximated to Gold...."

> *Lead is a Metallic Body, livid, earthy, ponderous, mute, partaking of a little Whiteness, with much paleness, refusing the Cineritium,[10] and Cement, easily extensible in all its dimensions, with small Compression, and very fusible without Ignition. Yet some Men say that Lead in its own Nature, is much approximated to Gold; these judge of things, not as they are in themselves, but according to sense, being void of Reason, and not conceiving the Truth.[11]*

Though he believed that lead did not bear gold in potential, he did not rule out the possibility that lead could be transmuted into another metal. In the very next paragraph, he related how lead could be converted into tin and even into silver.

> *It has much of an Earthy substance, and therefore is washed, and by a Lavament[12] converted into Tin, by which it appears, that Tin is more assimilated to the perfect. It is also by Calcination made Minimum; and by hanging over the Vapor of Vinegar, it is made Ceruse. And though it is not near to perfection, yet to our Art, we easily convert it into Silver, not keeping its own Weight in transmutation, but acquiring a new Weight, which it obtains by our Magistry.*

Pseudo-Geber had no doubt, then, that lead, through a series of operations, could be transmuted into another metal,[13] but not into gold. Saturn could eventually be coaxed into Jupiter or Venus (tin or copper, respectively), he believed, but not into Sol.

Pseudo-Geber's view on the nature and transmutation of lead, in contrast with the later Johan Hollandus's (see below), was in line with the ancient view that lead was a metal from which the other three base metals of tin, copper, and iron should originate.

> [In this view] . . . metals [were] born within the Earth-Mother, where they 'ripen[ed]' from the original embryonic state into the full perfection of silver and eventually gold. . . . The idea of ripening was crucial to alchemy; therefore, it was important to know the sequence of steps in such a process. Over time, sequences were proposed that began with the metal considered to be the least precious and ended with gold. In such sequences of metals . . . lead appears repeatedly in the first place as a trace of the ancient belief that it was a metal from which should originate the remaining three base metals (copper, tin, and iron).[14]

Basil Valentine

The correspondence of Saturn and lead takes a curious twist in Basil Valentine's (1394-1459) book *Of the Great Stone of the Ancients*.

> . . . the king's crown should be pure gold, and a chaste bride should be married to him. Take the ravenous gray wolf that on account of his name is subjected to bellicose Mars, but by birth is a child of old Saturn, and that lives in the valleys and mountains of the world and is possessed of great hunger. Throw the king's body before him that he may have his nourishment from it. And when he has devoured the king, then make a great fire and throw the wolf into it so that he burns up entirely; thus will the king be redeemed. If this is done thrice, then the lion has conquered the wolf, and nothing more to eat will be found on him; thus is our body completed at the start of our work.[15]

From *The Alchemy Collection: The Works of Basil Valentine*[16]

The reference to the "... child of Saturn ..." suggested a derivative associated with lead. In this case, as Lawrence M. Principe demonstrated in his *Secrets of Alchemy*, (2012), the association was with stibnite (antimony ore),[17] which was held to be closely associated with lead.[18] Stibnite was often found in the same deposits as Galena and Pyrite.

> ... *thus lead may be converted into silver, for if you place a pound of lead in the fire, it rectifies it and matures it, and most of it is burnt away, leaving a small part as silver. Now since it is possible for a part of the lead to be changed into silver there is nothing to hinder the conversion of the whole. In the same way, silver may be converted into gold, by the refinement of the smelting fire only.*[19]

Paracelsus

Though there were differences among alchemists over whether lead could be transmuted into gold or just into the base metals, there was widespread agreement that, in either case, lead was the basest material and could serve as the origin of other metals. Paracelsus (1493-1541) agreed with this premise. In his *The Economy of Minerals*, Chapter XXI, "Concerning Metals Free by Nature, Perfect and Imperfect; and First Concerning Saturn, or Lead," he said this about Saturn:

> *Saturn has obtained a body the blackest and densest of all (though white, yellow, and red inhere therein), Mercury a similar one, and Salt one above all others fusible. By corruption, it is easily reduced to its spirit, to white or yellow cerussa, to minium, and lastly, to glass, like the rest.*[20]

Here, lead is clearly understood to be the densest of all the metals, and the blackest, which may be a reference to a corrupt and impure state from which a purer metal might be obtained. Cerussa was also known as "Rust of Lead," or "White Lead," and could be either white or yellow in color. It was a powder or ash and was considered a poisonous body. In Rulandus's pharmacopeia, it was widely known as "White Lead."[21] Paracelsus also described Minium, known as red lead oxide (Pb_3O_4),[22] as Mercury of Saturn Precipitated or Saffron of Minium. In this same chapter, he understood lead as well as tin to be of a liquefiable sulphur, salt, and mercury, contrary to steel and iron.

Symbol for Minimum (Red Lead Oxide)[23]

Johan Hollandus

Johan Hollandus (Flemish physician, late 16th to early 17th century) took Rulandus's description a step further. Where Rulandus's lexicon affirmed that because of lead's internal nature it could be transmuted to tin, Hollandus claimed that lead could be perfected into gold. The notion that lead contained, in potential, the gold that so many alchemists sought, was explicitly averred in Hollandus's *The Mineral Work*, Chapter LI:

♄ [A variation of the symbol for Saturn] *is fine gold in its innermost, but impure. If its impurity were taken away from it and its innermost turned outside, it would be perfect gold.*[24]

In the same work, in *Chapter LXXVIII, The Twenty-First Work*, Hollandus reiterated this idea. Here he detailed the recipe for the production of the Philosopher's Stone from lead:

> *My son shall know how he can prepare the Philosopher's Stone from lead alone, without any other additions. It is as powerful as the one made of gold, all by itself, without any ferment, for lead is good gold in its innermost, and it lacks nothing except that it is impure and its innermost is not turned outside. If its impurity were taken from it and its innermost were outside, it would be good gold. All philosophers concur in this who have investigated the work and have found the truth. For all these reasons, no other ferment has to be added to* ♄ *in order to make the Stone.*[25]

Thus, in contrast to Pseudo-Geber, Hollandus considered lead to actually be "fine gold" or "good gold" in potential. One only had to remove the impurities and bring forth the innermost or true nature of the lead. (A complete analysis comparing Pseudo-Geber recipes to those of Holland's–to tease out further insights into the lead or the Saturn of the alchemists–would be an interesting exercise, but it is beyond the scope of this essay.)

George Starkey
George Starkey (1628-1665), the Colonial American alchemist, applied the term "Saturn of Antimony" to the result of the fusion of two reguluses. In an experiment in 1653, he prepared a regulus by fusing stellate regulus (metallic antimony) with colcothar, a copper-containing residue left from the preparation of *aqua fortis* from niter and vitriol. He then prepared a second regulus by fusing stellate regulus with minium (red lead oxide). When he fused these reguluses together, he came up with lead, which he called "Saturn of Antimony."[26]

The recipes of the alchemists, including those recounted above, often involved processing lead or analyzing processes that incorporated lead at some stage. Observations of these processes led to additional descriptions and terms for lead, lead ores, and lead compounds, such as Manufactured Plumbago (see below). Cupellation, one of these processes, involved the heating of various substances, including lead.

Cupellation

Speculation that lead could be transmuted into other metals was supported by observations of cupellation, which was viewed as a purification process. Cupellation has been known since about 2000 BC. The Greeks used it for the purification of silver, which was often found in lead ores. When the lead was heated–during the cupellation process–silver was often separated out. This led to speculation on the part of early alchemists that at least some of the lead was transmuted into silver. Evidence of how this conclusion might have been reached was found in the writings of the 13th-century Muslim alchemist al-Iraqi (full name: Abu al-Qasim Ahmad ibn Muhammad al-Iraqi al-Simawi), who was the author of *The Book of Acquired Knowledge concerning the Cultivation of Gold*: "... thus Lead may be converted into Silver, for if you place a pound of Lead in the fire, it rectifies it and matures it, and most of it is burnt away, leaving a small part as Silver."[27]

Grey, Black, and White

According to Rulandus, lead manifested in three qualities—gray, black, and white. Grey lead, known as *Plumbum Cinereum*, was actually bismuth, which was known as a metal in its own right by the miners of the time. It was considered nobler than lead but inferior to silver, and its ore was similar to Galena but was not brittle like Galena. The bismuth ore was described as blacker than rude, lead-colored silver, but most alchemists believed that it, nevertheless, contained silver. This ore was also referred to as the Roof of Silver since, in the mining of the ore, silver would usually be found at a deeper level. Other names for *Plumbum Cinereum* included the White Flower of Grey Lead, Aldeburg Lead, Crustaceous Lead from the Valley of Joachim, and Glebous Lead of Schneberg[28] (Presumably Aldeburg, the Valley of Joachim, Glebous, and Schneberg indicated locations at which the ore was found).

Black Lead, *Plumbum Nigrum*, was extracted from Galena (lead sulfide, PbS) or from Pyrites (iron II disulfide, FeS_2), according to Rulandus.[29] The reference to pyrites may be confusing, but by the mid-1500s, pyrites had become a generic term for all sulfide minerals and was associated with other sulfides and oxides in quartz veins, sedimentary and metamorphic rock, and coal beds.[30] (Lead appears in small quantities in pyrite from unmetamorphosed sedimentary rocks.)[31]

In Spain, Black Lead was known as Galena, which Rulandus referred to as Glance. The Greeks, however, because of its lead color, called Black Lead Molybdenum (The Greek word for lead), while the Romans referred to Galena as Plumbago or the Stone from which lead is made. Black lead, therefore, came to be known as the Lead Stone, Vein of Lead Ore, or Lead Earth. Other descriptions included Leaden Stone, Washed Lead, and Lead Scoria. Though Galena might be described as having various colors—such as Black, Blue, Yellow, or "Liver-colored" by various authors—there is no difference in its chemical make-up—only in its appearance. According to Rulandus, these variations in color are attributed to "... different gasses stirred up from the bowels of the earth ... [or] ... tempered by the heat of the earth." Presumably, he was referring to discolorations caused by heat and/or reactions with various substances.[32]

Plumbago was distinguished in two ways—native and manufactured. (The native form is discussed above.) Manufactured Plumbago was produced in furnaces; it went by various names. In German texts it was known as Graphite or Compressed Galena. Others called it Silver Litharge, or—in the case of Pliny— a species of Silver Litharge,[33] because of its affinity to Scoria of Silver (Scoria was Refuse or Coarse Matter).

The descriptions of manufactured Plumbago, however, gave rise to confusion and disagreements (we should not be surprised!). Matthaeus Sylvaticus, in his *Pandectist* (1317),[34] used the term Molybdena and called it Burnt Lead and the Refuse of Lead. However, according to Rulandus, this molybdenum or Black Lead was not lead, nor the Refuse or Scoria of Lead. He believed it was a Residuum which contained silver and lead and adhered to the sides of furnaces.[35]

White Lead, *Plumbum Candidum*, was often called Ceruse or tin, but it was not really tin, which Rulandus rightly pointed out in his lexicon. White Lead was purer and more perfect than Black Lead. Pliny referred to it as Cassiterion—a reference to the Cassiterides Islands. Like Black Lead, it was extracted from Galena and pyrites.[36] White Lead occurred in mineral form as a hydrate of cerussite ($PbCO_3$, the hydrate being $2PbCO_3 \cdot Pb(OH)_2$).

Ceruse, also called Cerusa or Cerussa, had various descriptions. According to Rulandus, Cerusa was known as Rust of Lead, White Lead, the Psimytim, Position, or Aphidegi of the Greeks, and, according to Dioscorides[37] and Nicander,[38] a Poisonous Body. Rulandus described Cerusa as the powder, ash, rust, or ceruse of White or Black Lead.[39]

CHAPTER 4

Lead Compounds

In studying alchemical and chymical texts, therefore, the reader does not only have to contend with the plethora of names for lead and lead ores. The names for lead compounds created by the alchemists were just as numerous and confusing. Litharge and minimum, the two compounds mentioned above, resulted from the processing involved with manufactured plumbago. Litharge was a reddish-yellow crystalline form of lead monoxide, formed by fusing and powdering massicot (a yellow powder form of lead monoxide). Minium, a scarlet crystalline powder, was formed by roasting litharge in air.

Symbol for Lead Monoxide[40]

Litharge was also spoken of in different ways. For example, litharge of gold was litharge mixed with red lead; litharge of bismuth was the result of the oxidation of bismuth; and litharge of silver was a by-product of separating silver from lead, which resulted from mixing silver ore, litharge (crude lead oxide) flux, and charcoal, and then smelting the result in clay and stone furnaces.[41]

Various Symbols for Litharge[42]

In addition to the by-products of these processes, other lead compounds, known at the time, included Naples yellow or Cassel yellow—an oxychloride of lead, made by heating litharge with sal ammoniac (ammonium chloride); Venetian ceruse[43] (also known as Spirits of Saturn)—a mixture of water, vinegar, and white lead; and sugar of lead, which was lead acetate (also known as sugar of Saturn)—formed by dissolving lead oxide in vinegar.[44] According to the *Summa*, sugar of lead could be further used to obtain metallic lead. By dissolving zinc in a solution of lead acetate and acetic acid, lead could be deposited as thin, crystalline leaflets known as Arbor Saturni or Tree of Saturn.[45]

Symbol for Sugar of Lead (Lead Acetate)[46]

Conclusion

Though considered the basest of the seven metals known to the ancient and medieval alchemists, lead occupied a significant and influential function. The various descriptions and objectives offered in their writings to reveal lead's role attested to this "weighty" position. In addition, whether the alchemists' goals were the attainment of the Philosopher's Stone, the transmutation of a base metal into gold or silver, or the more mundane production of particular substances, the starting point was often lead. Like slow-moving Saturn whose orbit encompassed the orbits of the more prominent planets, lead's orbit in the alchemical heavens encompassed the other six venerated metals and numerous derivative substances. Lead, the least noble of metals, was, in many ways, the most preeminent.

Notes

1. Tom Jackson, *The Elements: An Illustrated History of the Periodic Table* (New York: Shelter Harbor Press).
2. Cronus learned from Gaia and Uranus that he was destined to be overcome by his own children, just as he had overthrown his father. As a result, although he sired the gods Demeter, Hestia, Hera, Hades, and Poseidon by Rhea, he devoured them all as soon as they were born to prevent the prophecy from coming true.
3. The English word "plumbing" was derived from the Latin for Lead (Plumbum).
4. Martinus Rulandus, *A Lexicon of Alchemy or Alchemical Dictionary, Containing a Full and Plain Explanation of All Obscure Words, Hermetic Subjects, and Arcane Phrases of Paracelsus* (By the care and expense of Zachariah Palthenus, Bookseller: Frankfurt: 1612), 1.
5. Rulandus's lexicon was purported to be a summary of all the technical terms and descriptions from the Hermetic tradition and used by Paracelsus. The reader should bear in mind that Paracelsus died some 70 years earlier. It is possible that these terms came not only from Paracelsus, but his followers as well. We cannot simply attribute these to Paracelsus alone. Also, the reference to the Hermetic tradition in Rulandus's lexicon would imply that at least some of these terms came from the alchemical tradition pre-dating Paracelsus, but were then used by Paracelsus or his followers.
6. Livid—having a discolored, bluish, or gray-bluish appearance; Terrene—earthy.
7. Possibly a reference to cupellation, considered to be a process of purification.
8. Referred to as such because it was seen as possessing a soul.
9. Rulandus, 250.
10. Possibly a sand bath. Also known as *a qadr* in Arabic alchemy, or balneum siccum, balneum cineritium, or balneum arenosum in Latin alchemy. Retrieved from Wikipedia 21 Jan 2023.
11. Geber, *Summa Perfectionis, The R.A.M.S. Library of Alchemy Volume 9,* (Stuarts Draft, VA, The RAMS Publishing Co: 2015), 40.
12. A washing or cleansing.
13. Geber, 40.
14. Vladimir Karpenko, "Systems of Metals in Alchemy," *Ambix*, Vol. 50, Part 2, July 2003, 216.

15. A more complete explanation of this excerpt can be found in the essay "The Masks of Antimony."
16. Basil Valentine, *The Alchemy Collection: The Works of Basil Valentine: Including The Triumphant Chariot of Antimony, The Twelve Keys, and Of Natural & Supernatural Things, 3rd edition*, Adam Goldsmith, editor (Vitriol Publishing: 2013).
17. Lawrence M. Principe, *The Secrets of Alchemy* (Chicago, University of Chicago Press: 2013), 146.
18. Pliny, Historia naturalis, bk 33, sect. 34.
19. Vladimir Karpenko, "The Chemistry and Metallurgy of Transmutation," *Ambix*, Vol. 39, Part 2, July 1992, 47-62. See also, Abu'l-Quasim. Muhammad ibn Ahmad AI-Iraqi, "Book of Knowledge Acquired Concerning the Cultivation of Gold", trans. by E.J. Holmyard, *Hamdard Medicus*, XX(1977), 19.
20. Arthur Edward Waite, ed., *The Hermetic and Alchemical Writings of Aureolus, Philippus Theophrastus, Bombast, of Hohenheim, called Paracelsus the Great*, (Mansfield Centre, CT, Martino Publishing: 2009), 110-111.
21. Rulandus, 96, 232.
22. Formed by roasting litharge in air. Scarlet crystalline powder.
23. Philip Wheeler, *Alchemical Symbols, 4th edition, R.A.M.S. Library of Alchemy Vol. 21*, (Kansas City, MO, R.A.M.S. Publishing Co.: 2018), 49.
24. Johan Hollandus, *The Mineral Work, The R.A.M.S. Library of Alchemy Volume 26*, (Stuarts Draft, VA, The RAMS Publishing Co: 2015), 91.
25. Ibid, 136-137.
26. William R. Newman, and Lawrence M. Principe, *Alchemy Tried in the Fire: Starkey, Boyle, and the Fate of Helmontian Chymistry*, (Chicago, The University of Chicago Press: 2005), 130-131. Starkey did not observe differences between his "Saturn of antimony" and "natural lead." See notes 88 and 89 on page 131 of the above reference.
27. Karpenko, "Systems of Metals in Alchemy," 220. See also, Vladimir Karpenko (1992) "The Chemistry and Metallurgy of Transmutation," *Ambix*, 39:2, 47-62; and Vladimir Karpenko, "Transmutation: The Roots of the Dream," *Journal of Chemical Education*, 72 (1995): 338-385.
28. Rulandus, 250-254.
29. Rulandus, 250-251.

30. "De re metallica," *The Mining Magazine*, Translated by Hoover, H.C.; Hoover, L.H. London: Dover. 1950 [1912]. See footnote on p. 112.
31. M. Wampler and J. L. Kulp, "An isotopic study of lead in sedimentary pyrite," *Geochimica et Cosmochimica Acta* Volume 28, Issue 9, September 1964, Pages 1419-1458. Retrieved from Science Direct website 01 Jan 2023. https://www.sciencedirect.com/science/article/abs/pii/0016703764901607.
32. Rulandus, 254-255.
33. Litharge, a secondary mineral, forms from the oxidation of galena ores as coatings and encrustations—soft, red, greasy-appearing crusts. It has a tetragonal crystal structure and is dimorphous with the orthorhombic form massicot (PbO). See Wikipedia, N.N. Greenwood, A. Earnshaw, Chemistry of Elements, 2nd edition, Butterworth-Heinemann, 1997.
34. A medieval Latin medical writer and botanist (c. 1280 – c. 1342). He authored an encyclopedia of medicines (1317) entitled *Pandectarum Medicinae*. See The Encyclopedia of Medicaments, Library of Congress. https://www.loc.gov/item/2021666857/ Retrieved 27 Jan 2023.
35. Rulandus, 255-256.
36. Rulandus, 251-253.
37. A Greek physician (40-90), pharmacologist, botanist, and author of *De materia medica*.
38. Greek poet, physician, and grammarian (2nd century BC).
39. Rulandus, 96.
40. Wheeler, *Alchemical Symbols*, 49.
41. Litharge. Wikipedia. Alan Probert, "Bartolomé de Medina: The Patio Process and the Sixteenth Century Silver Crisis" in Bakewell, Peter, ed., *Mines of Silver and Gold in the Americas*, (Variorum: Brookfield, 1997), 96.
42. Wheeler, *Alchemical Symbols*, 51.
43. A 16th-century cosmetic used as a skin whitener. Most famously used by Elizabeth II.
44. The Alchemy Website. http://levity.com/alchemy/. Retrieved 01 Jan 2023.
45. Karpenko, "Systems of Metals in Alchemy," 222.
46. Wheeler, *Alchemical Symbols*, 49.

Chapter 5

Tin, Jupiter's Alchemical Influence

Introduction
From ancient times, tin[1] has been an essential metal in bronzes, brasses, and pewter. Its occurrence in the earth—2 parts per million, however, is relatively rare compared to other metals such as iron (50,000 ppm), copper (70 ppm), lead (16 ppm), and arsenic (5 ppm). Only silver with 0.1 ppm and gold with 0.005 ppm are rarer.[2] Consequently, ancient sources of tin were extremely scarce, and the metal usually had to be traded over very long distances to meet demand in areas which lacked tin deposits. Known sources of tin in antiquity included the southeastern tin belt that runs from Yunnan in China to the Malay Peninsula; Cornwall and Devon in Britain; Brittany in France; the border between Germany and the Czech Republic; Spain; Portugal; Italy; and central and South Africa.[3] Tin was of interest to the alchemists, metallurgists, and naturalists of antiquity and the Middle Ages. For the alchemists, in particular, tin was believed to be similar to silver, but in an impure state. Acquiring knowledge of tin, then, was believed crucial to the generation of silver and to a medial stage in the transmutation to gold, since they also believed silver was close to gold.

The following is a survey of their interest; especially in tin's "Whitening" ability (i.e., the tendency to render other metals silvery) and its peculiar creaking sound when the metal was deformed. First, however, a little technical background is in order.

Twinning of Tin
Artisans valued tin for its usefulness in making utensils, tools, and weapons. Physicians sought tin compounds for medicinal treatments. Alchemists, however,

sought to transmute tin into silver or to produce tin compounds–Elixirs that could transmute less noble metals into silver or, possibly, gold. An interesting characteristic exhibited by tin, whether it was being worked by artisans, physicians, or alchemists, was a peculiar noise made when it was stressed. Described variously as a "cry," a "shriek," or a "creak," this sound was produced by a crystalline phenomenon known as crystalline twinning.

Crystalline twinning within a mineral is the result of a phenomenon that causes a symmetrical intergrowth between two or more adjacent crystals, creating a tight bond between the crystals. The crystals are oriented in such a way that they share some of the same lattice points in a symmetrical manner. The surface along which the lattice points are shared is called a twin plane which forms a boundary that is a mirror image misorientation of the crystal structure. Twinning occurs when a shear force, acting along the twin boundary, causes the atoms to shift out of position.[4] As the twinning occurs, a sound is emitted which is perhaps more accurately described as a "soft whine" than the alchemists' "cry" or "shriek."

Twinning: Applied stresses to the crystalline structure (figure on the left) may cause the displacement of the atoms, resulting in the formation of a twin (figure on the right).[5]

Alchemists believed the "cry" to be an imperfection of tin, and if tin could be cured of its "flabbiness, cry, blackness, and odor," it would be transformed into silver. They discovered that corrupting tin with impurities was one way to eliminate its "cry." Unfortunately, this did not get them any closer to silver.[6]

Pliny the Elder
Pliny, the Roman naturalist, recorded that according to the Greek historian Timaeus (born 356 or 350 BC, died c. 250 BC), tin could be found on the island

of Mictis (or Ictis, an island off the coast of England involved in tin-trading) within six days' sailing from Britain. The Britons would cross in coracles constructed of osier covered with sewn hides to recover the tin. Pliny also recorded several other locations in which tin was found, including the Pyrenees and islands opposite Celtiberia that the Greeks called Cassiterides (Greek, meaning "Tin Islands", from κασσίτερος, *kassíteros*, "Tin") because of the rich supply of tin found there.[7]

In addition to identifying the location and occurrence of tin, Pliny described the metal's characteristics, its use in making bronze, and assay techniques. In the time of Pliny, tin was thought to be a type of lead. As Pliny explains, there were two kinds of lead, black and white (*Naturalis Historia*, XXXIV, 47). The white, which the Greeks called cassiteros (an ore of tin), was the most valuable and was obtained in Lusitania and Gallaecia—found on the surface of the earth from black-colored sand. Because of its greater weight, the ore was able to be separated from the sand and small pebbles in the dried beds of torrents. At the time, tin was detected only by its weight. Tiny pebbles of tin would appear sporadically, especially in the dried-up beds of what were once fast-moving rivers. Workers would wash the sand and heat the residue in furnaces. Tin was also found in gold mines called alutiae, through which a current of water was sent to wash out the black pebbles with their white spots of tin. Since the pebbles weighed the same as the gold, they remained in the bowls that collected the gold. Afterwards, the pebbles were separated in the furnaces and fused into tin (*Naturalis Historia*, XXXIV, 157).[8]

Finally, Pliny described the use of tin to alloy copper in the making of bronze statues. The composition of bronze for statues, as well as for sheets of metal, was as follows: "... the ore is melted and to the melt is added a third part of copper scrap—that is, used, second-hand copper. This scrap contained an intrinsic, seasoned brightness, since it was subdued by friction and tamed by use. Tin was also alloyed with it, in the proportion of one part of tin to eight parts of copper (*Naturalis Historia*, XXXIV, 97)."[9]

Symbol for Tin or Jupiter[10]

CHAPTER 5

Pseudo-Democritus (First Century CE)
Tin played a central role in Pseudo-Democritus's recipe for making silver. In his treatise "On the Making of Silver," he explained that first a volatile substance had to be made:

> *Take the mercury that comes from orpiment, or from realgar, or according to your knowledge, and make it solid as is customary; lay it on the copper or on the iron which have been purified with sulfur, and they will turn white. Whitened magnesia produces the same effect as well, and orpiment that has been turned inside out, and roasted cadmia, and unburnt realgar, and whitened pyrite, and white lead roasted together with sulfur. You will melt iron by adding magnesia, or half of sulfur, or a pinch of magnetite. For magnetite has affinity with iron. Nature delights in nature.*

The next step involved the addition of tin:

> *Take the above-mentioned volatile substance and boil it with castor or radish oil, mixing in a pinch of magnesia. Then take tin and purify it as is customary, with sulfur, or with pyrites, or according to your knowledge. Pour it over the volatile substance and make a mixture. Let it be roasted in an enveloping fire, and you will find it similar to white lead. This drug makes any (metallic) body white....*[11]

The result of this recipe, then, would be a metallic white body, a reference to silver. What formed was a coat of tin oxide over the surface of the metal, which had a whitish to greyish-white color. The reference to "white lead" may have been a reference to cerussite, a mineral of lead ($2PbCO_3 \cdot Pb(OH)_2$). Because of the similarities, the term "white lead" was used for tin.

Pseudo-Democritus did not offer an explanation for the "creaking" phenomenon of tin. But he did offer a remedy that would remove tin's characteristic sound in a recipe that described the uses of sulfur for whitening copper, softening iron, silencing tin, etc.:

> *Take white sulfur; you shall whiten it, dissolving it with urine in the sun or alum and salt brine: it will shine having become totally white. Dissolve it with realgar or heifer's urine for six days, until the*

> *pharmakon ["the medicine"] nearly approaches the likeness of marble; when it becomes so, great is the mystery for it whitens copper, softens iron, takes away the creaking of tin, makes lead not fusible (takes away the fluidity of lead), and makes substances unbreakable and dyes permanent. For sulfur mixed with sulfur makes substances divine, since they (sc. sulfurs) have a close kinship with each other. Natures rejoice with nature.*[12]

The recipe described a preliminary treatment of sulfur to whiten it, using urine, alum, or brine. The next step was to melt sulfur with realgar (an arsenic sulfide) or dissolve it in urine until the mixture would become like marble. This would imply that a hardening or solidification operation was employed. Pseudo-Democritus used the term "pharmakon" to refer to a liquid substance. Thus, "... approaching the likeness of marble ..." may have referred to obtaining the color of marble. Either way, the final state of the compound involved a white solution of sulfur. He did not specify what this solution may have been, but he discovered that tin, hardened by sulfur, no longer produced its peculiar "cry,"[13]

From the recipe above, Pseudo-Democritus understood that a single ingredient (in this case sulfur) had different effects depending on the kind of metal to which it was added. Sulfur, as noted, would whiten copper, soften iron, or remove tin's cry.

This recipe also revealed a glimpse into his understanding of the nature of tin—unlike other alchemists, he did not consider tin as arising from lead. Even though the vocabulary of the recipe (e.g., "For sulfur mixed with sulfur makes substances divine, since they ... have a close kinship with each other ... (and) ... Natures rejoice with natures ...") is related to the theory of natural sympathy (According to this theory, substances that share similar properties or qualities are attracted or *sympathetic* to each other, while those with opposing properties repel or are *antipathetic* to each other), his observations argued against transmutation as a process involving natural sympathy, thus rejecting tin as a derivative of lead. Natural sympathy conceived transmutation as beginning with a mixture of reactive ingredients, with or without reference to a prime matter. This first step, known as "blackening," consisted in obtaining an unqualified substrate, understood as the fusion of a metal into a dark mass, and then mixing this substrate with a substance or substances chosen for sympathetic reactivity. The notion that sulfur would have different effects on various metals ruled out that molten metal was a prime matter or unqualified substrate.

Instead, the different results produced by the mixing of a single substance to different metals implied that the molten metals had different qualities. This understanding was contrary to another theory as well. Common at the time, lead was believed to be the most basic of metals and that all substances were made of lead, since the three other bodies of copper, iron, and tin were believed to be derived from lead. Dissenting from this view, Pseudo-Democritus held that copper, iron, and tin were unique substances, but collectively, including lead, were called the four bodies.[14]

Tin Symbol[15]

Pseudo-Khalid Ibn Yazid
Another recipe in which tin was used for "whitening" (the color alchemists used to describe silver) was provided by the author of *Liber secretorum artis* (*Book of the Secrets of the Art*) or *Liber secretorum alchimiae* (*Book of the Secrets of Alchemy*). This treatise has been attributed to Khalid ibn Yazid (c. 668-704), one of the sons of the caliph Yazid I (the true author and date cannot be ascertained). The recipe described how "the stone" could be used to transmute base metals into silver—Chapter XV, entitled "The Manner of Operating the Stone for Whitening":

> *. . . take 250 drachmas of lead or tin, and melt it. Then, when it has become liquid, throw upon it one drachma of cinnabar, that is, of this medicine which you have brought to this honored state and to this high order, and let that tin or lead be retained lest it fly away [volatilization] from the fire, and it will whiten it, and will extract its detriment from it, and its blackness [impurities], and convert it into a perpetual tincture. Then take one drachma of those 250, and throw it upon 250 [drachmas] of tin, or brass, or copper, and it will convert it into silver better than the mineral ore . . .*[16]

Cinnabar (mercury sulfide, HgS) was used to remove impurities from either tin or lead and produce a tincture that could "whiten" tin, brass, or copper, converting the metal into silver. The products formed from the reactions of tin or lead with mercury sulfide would have been amalgams of HgSn and HgPb. In regard to tin, the amalgam obtained is a soft, silvery paste from which mercury could be easily recovered upon heating.[17] This amalgam of tin and mercury, because of its silvery color, has also been known as "Mercury of Tin" (e.g., Zosimos).[18] As a paste, then, the amalgam could easily be joined to a metal, imparting a silvery color and giving the appearance that the metal had been converted into silver.

Al-Razi (Abu Bakr Muhammad ibn Zakariya al-Razi, c. 865-923 CE)

In his discussion of making the Elixir, al-Razi described the use of tin and lead in the transmutation to silver or gold. By Elixir, al-Razi meant that medicine, which "... when fed onto a molten Body" would convert that body into silver or gold.[19] He provided a recipe from the "Sections on the Coagulation (Solidification) of Mercury":

> *Take of Mercury, as much as you want, and rub it with (powdered) mustard for 1 hour until it is black. Then heat it with vinegar and salt until it is purified (from the blackness). Next put it in a pit in the ground and pass over it a feather dipped in olive oil, until the surface is entirely covered with the oil. Then scatter over it a small amount of white ashes which have been passed through a sieve, and pour on it melted Lead (usrubb) or Tin (qala'i) till it is covered to the thickness of one finger. Do this repeatedly till it is coagulated into a stone. I mean, melt Two Leads (rasasayn) several times, and pour them over the Mercury. Better than this. Scatter over it, instead of ashes, White Marqashitha if you want it for the White and, if you want it for the Red, Golden Marqashitha. Pour over it for the White, Tin, and for the Red, Lead ... Also for the Red, Yellow Sulphur (is used), and for the White, Sal-Ammoniac.*[20]

Similar to Pseudo-Khalid, al-Razi described a recipe for the amalgamation of tin or lead with mercury. In his recipe, al-Razi also listed possible substances that could be used for "whitening," an amalgam of tin and mercury among them. Instead of the cinnabar used by Pseudo-Khalid, al-Razi used vinegar (acetic acid) and a salt (he does not specify the specific salt) to remove any impurities (blackness) from the mercury (more appropriately quicksilver). Once the mercury was purified, it could then be coagulated or solidified in several different ways: White Marqashita (marcasite—white iron pyrite, FeS_2), sal-ammoniac (this may have been alum),[21] and tin for "whitening;" or Golden Marqashita (golden marcasite), yellow sulphur (probably an arsenic sulfide), and lead for "reddening."[22]

Later, in the chapter "On the Sublimation of Mercury," al-Razi expanded on these two methods: one for "reddening" and one for "whitening." To sublimate mercury for "whitening," explained in the section entitled "Sublimation of Mercury for the White" (see below), al-Razi began with mercury that had been solidified with tin as described in the recipe above:

> *You take as much as you require of mercury that has been coagulated by the odor of tin, and bruise it with an equal quantity of white alum, the like quantity of salt, and the like quantity of ashes. You next sprinkle vinegar over it, after placing it on a salaya (flat stone mortar) and triturate it thoroughly for three hours a day, one hour in the morning, one hour at noon, and one hour in the evening. Then you place it in a flask (qarura) covered with clay. Close the head of the flask and place it on hot ashes in a tannur which has just been used for bread making. You leave it there for one night, and in the morning transfer the substance to the pot of the uthal, after again triturating it.*[23]

Al-Razi also provided a recipe in which a tin-mercury amalgam was amalgamated with silver, found in the third chapter entitled "The Calcination of Silver through Amalgamation":

> *Amalgamate one part tin with four parts mercury and grind it thoroughly. Then add three more dirhams of mercury, without washing it, then grind it one additional day, press it out and amalgamate the silver with that which was pressed out of it, three for one, and grind and roast it, until it has become a white calx. Add to it three times its weight dissolved mercury, let it solidify and add one dirham of it to tin amalgamated with iron, then you will find it as pure silver as God wills.*[24]

The silver-tin amalgam, as noted in the above recipe, was calcined to a white calx that was then mixed with mercury and applied to a tin-iron alloy to produce what was believed to be silver. The mixture of the mercury with the ground calx, in a ratio of three to one, would have been a paste and consisted primarily of tin oxide (SnO_2) and possibly some mercuric oxide. Silver would not have oxidized even at high temperatures. The actual presence of mercuric oxide, however, may be questionable for a couple of reasons. First, mercury oxidizes above 350°C, and so the question would be: Was the amalgam roasted to a high enough temperature to produce mercuric oxide? Also, mercuric oxide tends to be yellow or red in color and the result of a "white calx" would argue against the presence of mercuric oxide, but does support the presence of tin oxide since calcined tin is known to result in a white oxide. This paste, because of the added mercury, would have had a silvery-color and would coat the tin-iron alloy, causing it to appear as silver. In the mind of the alchemist this, conceivably, would have confirmed the belief that tin was an imperfect form of silver, and if properly prepared, could be made into an elixir for the transmutation of other metals into silver.

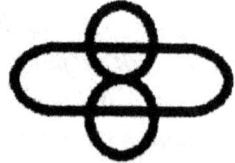

An Alchemical Symbol for Tin[25]

Avicenna (Ibn Sina, 980-1037 CE)

Avicenna's name is a Latin corruption of the Arabic patronym Ibn Sina. He was a Persian philosopher and physician born in Kharmaithan near Bukhara in what is known today as Uzbekistan. Over two hundred and fifty books have been attributed to him. The most important of these were *The Book of Healing*, a philosophical and scientific encyclopedia, and *The Canon of Medicine*, a medical encyclopedia. In addition to medicine, Avicenna was also interested in alchemy. He differed from many alchemists, however, believing, at least intuitively, that transmutation was impossible.[26]

Even though Avicenna did not believe transmutation was possible, he did provide descriptions of some operations that were representative of the methods employed by alchemists in Iraq and Persia in the tenth and eleventh centuries CE. For instance, in a treatise entitled *The Treatise of the Most Excellent of the Moderns, Al-Shaykhal-Ra'is Abu 'Ali al-Husaynibn 'Abdullah Ibn Sina Al-Bukhari-God Give Peace to His Tomb and Sanctify His Soul—for the Imam Abu 'Abdallahal Baraqi-God's Mercy on Him!-On the Sublime Art,* apparently written in response to a request, Avicenna provided a concise summary of the operations used in making the Elixir, derived from al-Razi's *Kitab al-Asrar* (see above). The preliminary operations can be summarized as follows:

> *(1) From a "Spirit" (in this case Sulphur), a Red Tincture was extracted by heating with a solution of Sulphur in "Strong Water" (Caustic Soda). The extract was evaporated to a solid and kept separately with the residual "Dregs". (2) Gold or Copper (but preferably Gold) was calcined with Sulphur. (3) Similarly, an amalgam of Mercury with either Tin or Lead was acted on by Sulphur. (4) The various products after separately undergoing sublimation, and conversion into wax-like incombustible substances, were dissolved, mixed, and the whole coagulated into what was believed to be the desired Elixir. . . .*[27]

This summary was followed by expanded treatments of each of these operations. Step three on the coagulation of mercury was as follows:

> *The Method of Coagulating Mercury for the Red*
> *It (the Mercury) is placed in a pit in the ground and its surface rubbed with a piece of wool saturated with oil. Then there is sprinkled over it Yellow Sulphur and Golden Marqashitha (Pyrites). Next, Tin (rasas) or Lead (anuk) is melted, and after being put aside until it has almost re-solidified, it is poured onto the surface of the Mercury. Notice if the latter has solidified; if not, the process is repeated as many times as is necessary until it (the mixture) coagulates into a Stone.*[28]

What is interesting in Avicenna's summary is that tin is included for making the Elixir for "reddening" when al-Razi was clear that tin was to be used for the Elixir for "whitening." Stapleton (Professor of Chemistry, Presidency College, Calcutta, in the early 1900s), et al., have pointed out that Avicenna may have haphazardly quoted al-Razi and others, and, furthermore, most likely did not carry out the experiments himself.[29]

Avicenna, however, did offer an explanation for the "cry of Jupiter":

> *As for Tin, it is probable that its Mercury is good but that its Sulphur is corrupt; and that the commingling of the two is not firm, but has taken place, so to speak, layer by layer, for which reason the metal "shrieks."*[30]

Here, Avicenna has asserted that the principles of Mercury and Sulfur were involved. In this case, the Mercury, that is, the fluid nature, is good. The Sulfur, the fixed nature, however, is corrupt or impure, preventing a stable, homogeneous mixture and resulting in a layering effect. For this reason, then, tin emits a noise (shrieks) when bent. Avicenna was close to the truth in identifying that a layering is involved. In actuality, when tin is subject to stresses that cause the layers of the metallic lattices to shift (see the discussion of twinning above), its characteristic "cry" is emitted.

Abufalah, the Saraqusti (11th century CE)

A Muslim Arab alchemist, Abufalah the Saraqusti, most likely lived in Syracuse, Sicily, during the eleventh century CE. His writings have been preserved in a Hebrew alchemical manuscript, Gershon ben Shlomo of Arles's (a Provençal scholar of the late 13th century) *Gate of Heaven*, a compilation of treatises summing up the sciences of the thirteenth century, published in 1547.[31]

In regard to tin, Abufalah wrote:

> *Tin is . . . near in its nature to silver, and differs from it in three respects: in smell, in softness, and in tone. And these features became attached to it at [the time of] its origin. Indeed, its softness is due to the abundance of its quicksilver, or due to the small measure of the cold which dried it or desiccated it; and the cause of its tone of its sound is the thickness of the sulphur that is in it; and its bad smell is due to the insufficiency of its ripening. Behold, through a known procedure it will become quicksilver, for it is close to its nature, as I shall reveal to you in the Gate that follows this, and due to a slight cause it will be saved from its traits, as we have said about silver, which is close to the nature of gold, and with thoughtful work we can transmute it into gold, as I shall reveal unto you likewise after this.*[32]

Abufalah made the same observation as many other alchemists would in regard to how similar tin was to silver. At the same time, however, he remarked on how tin differed from silver. Tin was softer than silver because of its greater amount of quicksilver (mercury). This understanding may have been derived from the Mercury-Sulphur theory of Jabir ibn Hayyan. (Since Abufalah disclaimed his writings were original and were a collection of older sources, it is possible that he was reiterating Jabir's theories.)[33] The sound (a reference to tin cry) was due to the Sulphur. Like the connection to the principle of Mercury in the first observation, a connection to the principle of Sulfur seems to be made in the second. If the tin was impure, contaminated with a sulfur compound, then its "bad smell" is understandable. This could have also led to the previous deduction connecting sulfur to the peculiar sound tin makes.

Alchemical Symbol for Tin[34]

Pseudo-Geber (Paul of Taranto, 1290-1330)
In *Summa Perfectionis* Geber provided a description of tin:

> ... *Tin is a Metallick Body, white, not pure, livid and surrounding little, partaking of little Earthiness; possessing in its Root Harshness, Softness, and swiftness of Liquefaction, without Ignition, and not abiding the Cupel, or Cement, but Extensible under the Hammer. Therefore, Jupiter, among Bodies diminished from Perfection, is in the Radix of its Nature of Affinity to the more Perfect, viz. to Sol and Luna; more to Luna, but less to Sol, as shall be clearly declared in the following. Jupiter, because it receives much Whiteness from the Radix of its Generation, therefore it whitens all Bodies not White; yet its vice is, that it breaks every Body, but Saturn, and most pure Sol. And Jupiter adheres much to Sol and Luna, and therefore doth not easily recede from them, by Examen (or Tryal of Cupel).*[35]

Pseudo-Geber pointed out one of the distinguishing characteristics of tin—its ability to resist cupellation (a refining operation in which ores or alloys are treated to separate out noble metals) and cementation (a process by which one substance penetrates and changes another substance usually by layering the two materials and heating to just below the melting point of either substance). It is, however, very malleable. He also noted that tin melts below red heat—"... swiftness of Liquefaction without ignition...."

Further, in the chapter, "Of the Nature of Jupiter, or Tin," Geber described the composition of tin:

> *We say that if Sulphur in the Radix of the Commixtion shall be participating in small Fixation, White with Whiteness not pure, not overcoming, but overcome, commixed with Argent vive partly fixed, and partly not fixed, white and impure; from the Mixtion Tin must needs follow. The Probation of these you will find by Preparation. For when you calcine Tin, you find a Sulphurous stink to arise from it; which is a sign of Sulphur not fixed. And although it yield no Flame, you must not therefore think it fixed. For it gives no Flame, not by reason of Fixation, but by reason of the Superancy of Argent vive in the Commixtion, preserving from Combustion. Therefore, in Tin is proved a twofold Sulphureity, and also a twofold Substance of Agentive. One Sulphureity is less fixed, because in calcining it casts out a stink as Sulphur. The Experience of the Mixture is proved by the First. The other is proved to be more fixed, by the continuation of it in its Calx, in the Fire which it hath, and yet it stinks not. That there is also a twofold Substance of Argent vive in it, whereof one is not fixed, and the other fixed, is proved; because it makes a crashing noise before its Calcination, but after it hath been thrice calcined, that crashing is not; the reason of this is, because the fugitive Substance of its Argent vive, making that crashing, is flown away. That the fugitive Substance of Argent vive is a Cause making that Strider, or crashing, is proved by washing Lead with Argent vive. For if Lead be washed with Argent vive, and after its washing melted in the Fire not exceeding the Fire of its Fusion, with it will remain part of the Argent vive, which gives this Stridor to the Lead, and turns it into Tin.*[36]

Pseudo-Geber claimed that tin was composed of a fixed (nonvolatile) white sulfur and an unfixed (volatile) white sulfur, a fixed quicksilver, and an unfixed quicksilver. The tin that Pseudo-Geber referenced could not have been pure tin since, when calcined, tin does not give off a sulfurous odor. It does, however, react with sulfur to form a sulfide. Geber's tin was more likely contaminated with a complex sulfide compound. Minerals such as stannite and teallite, for instance, contain sulfide compounds—copper, iron, and tin in stannite and lead and tin in teallite.

During the initial calcination, tin emitted an odor of sulfur, which, according to Geber, revealed the presence of unfixed sulfur. The sulfur, however, gave no flame, not because of its fixed nature, but because of the supremacy of the quicksilver (mercury), which prevented the sulphur component from combusting. With repeated calcinations, the sulphur became less fixed since eventually, no odor was detected.

Pseudo-Geber asserted that the quicksilver also had a twofold nature. This was proven by the tin losing its "crashing noise." Before being calcined, the tin would emit its characteristic "cry." After repeated calcination, it no longer makes the "crashing noise." Something has been lost, or, according to Pseudo-Geber, "... the fugitive Substance of Argent vive ... is flown away." Pseudo-Geber concluded this because of a previous experiment that treated lead with quicksilver. The lead, washed with quicksilver, and heated to below its melting temperature, gained the very "cry" that the calcined tin lost. The loss of this "cry," according to Pseudo-Geber, meant that a volatile component escaped, which he deduced was unfixed quicksilver. He knew this because he could induce the same creak in lead by adding volatile quicksilver.[37] To Pseudo-Geber, this proved that lead was converted to tin.

Pseudo-Geber, however, did not stop at describing the nature of tin. He also wanted to determine the relative quantity of the sulphur and quicksilver:

> *And in it (Tin) is the Equality of Fixation of the two Things compounding, viz. of Argentvive and Sulphur; but not Equality of Quantity; because in the Commixtion, the Agentive overcomes. The sign of which is the Easiness of Ingress of Argentvive in its Nature into it. Therefore, if in it were not a greater Quantity of Argentvive, the same (taken in its Nature) would not easily adhere to it. Wherefore it adheres not to Mars, unless with most subtle Ingenuity; nor unto Venus, by reason of the paucity of Argentvive in it, in its Commixtion. And this is evident, because it adheres to Mars most difficultly, but to Venus more easily; because that hath a greater abundance of Mercury, than Mars. The sign of this, is the easie Fusion of one, but most difficult Fusion of the other.*[38]

He reasoned that the ready amalgamation of tin with quicksilver proved that tin was composed primarily of the principle of Mercury, and thus was similar, in nature, to gold. Copper and iron, on the other hand, because they amalgamate only with difficulty, must have contained less quicksilver than tin.[39]

Pseudo-Maimonides (15th Century)
According to Patai, several treatises on alchemy, attributed to Maimonides (Jewish philosopher and Torah scholar, 1138–1204 CE), were written by an unknown author most likely in the 15th century.[40] Pertinent to this discussion was one particular treatise in which the author believed tin could be transmuted into gold in addition to silver. Pseudo-Maimonides described such a recipe for transmuting tin into gold in the treatise "Epistle of Secrets by the Rambam [Maimonides] of Blessed Memory":

> *Take tin as much as you like, and melt it, and after the melting put it instantly into juice of lemons or juice of sour citron, then put it in melted fat, and then in cold water, then in the juice of onions, then in juice of garlic, and then in water in which allabi fruit was cooked, then melt it another time with its own weight of moon [i.e., silver], and it will be very good yellow [i.e., gold].*[41]

Another Alchemical Symbol for Tin[42]

Paracelsus (c. 1493 – 24 September 1541)
In the chapter "Concerning the Spirit of Jupiter" of his work *Concerning the Spirits of the Planets,* Paracelsus described the nature of tin:

> *Concerning the spirit of Jupiter, this should be known, that it is derived from the white and pale substance of fire, together with a nature of peculiar crepitation and fragility, not malleable like Mars. It, therefore, heats other metals and renders them capable of being broken with hammers. An example of this may be seen when it is joined*

> *with Luna, for it can scarcely be brought to its former malleability, except with the greatest labor. The same effect it produces in the bodies of metals, it will do the same in human bodies. In these, it corrodes the limbs with severe burnings and decay, so that they are completely cut off from their perfect workings and lose them, so that they are unable to fulfill the necessary requirements of Nature. Nevertheless, this spirit has in it the virtue of removing cancer, fistulas, and other similar ulcers. . . .*[43]

The opening statement ". . . that it (tin) is derived from the white and pale substance of fire . . ." is rather enigmatic. Could the "white and pale substance of fire . . ." be a reference to the pale bluish-white color tin gives off when exposed to a flame? Could the statement be a reference to the Aristotelian notion of Fire, alluding to Paracelsus's understanding of tin's elemental nature? The meaning of this sentence remains unclear. What Paracelsus noted more clearly were the characteristics of tin: its peculiar cracking sound (crepitation) and its fragility, which compared to iron, lacked the malleability of the latter. Presumably, he was referring to tin's softness and pliability, unlike iron, which can be worked, but only at much higher temperatures. Another observation he offered in regard to tin's interaction with other metals was tin's ability to embrittle silver when ". . . it is joined with Luna. . . ."

In line with the primary objective—medicinal treatments—of Paracelsus's investigations, he connects the effects of tin on other metallic bodies to the effects on the human body. The organic compounds of tin can be poisonous if ingested, but inorganic compounds are usually considered non-toxic (SnF_2, tin fluoride is used in fluoride toothpaste).[44] In his description, Paracelsus was, presumably, referring to organic compounds of tin since these compounds are able to penetrate biological membranes and disrupt metabolic processes.[45]

Paracelsus also provided further details regarding tin in *The Third Book of the Archidoxies*, "On the Separations of the Elements from Metals":

> *Take salt nitre, vitriol, and alum, in equal parts, which you will distill into aqua fortis. Pour this water again on its feces, and distill it again in glass. In this aqua fortis, clarify silver, and afterwards dissolve it in it sal ammoniac. Having done this, take a metal reduced into thin plates in the same way, pour it on again, and repeat this until there be found at the bottom an oil, namely, from the Sun, or*

> *Gold, of a light red color; of the Moon, a light blue; of Mars, red and very dark; of Mercury, white; of Saturn, livid and leaden; of Venus, bright green; and of Jupiter, yellow.*
>
> *All metals are not thus reduced to an oil, except those which have been previously prepared. For instance, Mercury must be sublimated; Saturn calcined; Venus florified; Iron must be reduced to a crocus; Jupiter must be reverberated; but the Sun and Moon easily yield themselves.*[46]

His discussion begins with a preparation of *aqua fortis* (nitric acid) which he advised to purify with silver ("... in this aqua fortis, clarify silver ..."). The treatment with silver removed the hydrochloric acid present in the *aqua fortis* from any chloride impurities. If this was not performed in advance, silver chloride would form during the use of the acid. This was a well-known operation, a thorough explanation for which was provided by Vannoccio Biringuccio in *Pirotechnia*.[47]

After the silver chloride was removed, sal ammoniac was added to produce *aqua regia* (a mixture of hydrochloric and nitric acids) into which the metals were placed in either granular form or as hammered-out thin sheets. Paracelsus quickly pointed out, however, that not all metals could be reduced to an oil with this treatment unless they were properly prepared beforehand. Thus, mercury was to be sublimed; lead calcined; copper made into Flowers of Copper; iron made into its crocus; and tin reverberated.[48] Gold and silver, however, did not need such preparation as they would readily dissolve in *aqua regia*. The instruction to reverberate tin is confusing, since tin will dissolve in *aqua regia*, but the stannic oxide that forms from the reverberation of the metal will only dissolve in concentrated sulfuric acid.

By treating the metals according to the above recipe, Paracelsus observed that the dissolution of each metal yielded a compound with a distinctive color. For instance, gold (Sun) would form auric chloride, which has a red color; silver (Moon) would form silver nitrate and silver chloride, which are colorless, but silver was often contaminated with copper, and this impurity could explain the light blue color observed by Paracelsus; copper (Venus) would form copper chloride, which is green; iron (Mars) would form ferric chloride, which is brownish; mercury would form colorless compounds; and the lead (Saturn) compounds formed would most likely have been gray.

Tin, however, was the problem child. Most likely, a mixture of products would have been formed, including metastannic acid, stannic chloride (also known as Butter of Tin), and stannous chloride, all of which would have been colorless if pure. The presence of these compounds would not explain the yellow color unless the tin was impure. As mentioned in the discussion of Pseudo-Geber above, tin was often contaminated with other metals such as lead, copper, and iron. Possibly, the presence of an impurity would explain the yellow tint observed.[49]

After the metal had been dissolved, Paracelsus instructed that the "oil" obtained was to have fresh *aqua fortis* added to it, and then it was to be set in horse-dung (This would keep the temperature of the mixture slightly above air temperature), allowing the solution to clarify:

> *. . . distill it entirely with a slow fire, that the matter may be condensed at the bottom. And if the aqua fortis which ascends be distilled by a bath in this manner, you will find two elements together. But the same elements will not be left by all metals alike. For from Gold there remain in the bath earth and water; but air is in all the other three, and the element of fire remains at the bottom, because the substance and the tangibility of Gold have been coagulated by the fire; therefore, the substances will agree in its substantiality. From the Moon there will remain at the bottom the element of water, and in the bath the elements of earth and fire. For from the cold and the moisture is produced the substance and corporality of the Moon, which is, indeed, of a fixed nature, and cannot be elevated. From Mercury there remains fire at the bottom, and earth and water are elevated upwards. From Venus there also remains fire, and both, that is to say, earth and water, remain in the bath. From Saturn there remains the element of earth at the bottom, while fire and water are held in the bath. From Jupiter air remains at the bottom, while fire, water, and earth are elevated therefrom.*[50]

The two elements, understood in the context of the recipe, were the solid collected in the bottom of the vessel and the liquid *aqua fortis* collected in a connected receiver from which the acid solution could be redistilled. The natures of the two elements, however, were not alike depending on the metal dissolved. Paracelsus used the Aristotelian theory of the four elements of matter—Air, Fire, Earth, and Water—to describe the nature of each of the metals. For instance, from gold, the elements of Earth, Air, and Water remained in solution while Fire went to the bottom of the vessel. Apparently, Paracelsus distinguished which elements were present in the distillate by a process of elimination.[51] The element of Fire was, presumably, deduced by the color of the precipitate which could have been either gold (gold could be separated from the chloride by a more intense heat) or auric chloride ($AuCl_3$), both of which would have colors associated with fire—the color of auric chloride could range from red to golden yellow, depending on whether the anhydrous state or the monohydrate was present,[52] and the color of gold was likened to that of the Sun. Thus, the element of Fire was in the precipitate, and the other three elements would be left in the distillate. In the case of silver, Paracelsus concluded that Water was the constituent element since the product formed was, most likely, a fused silver nitrate or a concentrated solution of silver nitrate, both of which would appear as water. The elements he would have assigned to the distillate, then, would have been Fire, Earth, and Air. He followed this deductive process for the remainder of the elements.

In each case, Air was included with the remaining elements in the distillate. Tin, however, was the exception. Paracelsus concluded that what remained at the bottom of the vessel was corporeal Air. If what formed was stannic chloride ($SnCl_4$), then Paracelsus's conclusion may be reasonable. The fuming nature of stannic chloride and its tendency to easily boil (Stannic chloride's boiling point is 114.15°C)[53] would suggest the quality of Air.[54]

Vannoccio Biringuccio (c. 1480 – c. 1539)
Biringuccio, the Metallurgist, has provided a much more lucid description of tin than the alchemists thus far discussed. In the beginning of the fifth chapter, "Concerning Tin and Its Ore," of his work, *Pirotechnia,* he provided an observation that provides clarity to the enigmatic description of Paracelsus above:

> *Whoever has occasion to judge Tin in its whiteness from the testimony of his eyes alone would surely believe it to be purest silver, or something that comes very near to its nature. This is even more so when, on handling, it is found to be harder than Lead, to which it can be said to have a greater and closer resemblance. But whoever has experience with Tin knows that none of the other metals is less similar to it than silver. For Silver mixes with every metal, as does Gold; and they unite together, and the others do likewise; and except for their color, they change their natures but little. But this metal Tin,*

> *wherever it is found, acts like a powerful poison which envenoms and corrupts the others; and it does this not only with large quantities but also with every small amount; indeed, just the trace remaining where it has been melted is enough to embrittle silver and gold, as well as iron and copper.*[55]

Biringuccio has pointed out that the "whiteness" refers to the silvery color of tin, but then he quickly pointed out that in its nature tin was more similar to lead than silver, although significantly harder. That is, tin's ductility and malleability are closer to those properties of lead than to the other metals, especially silver. What makes the visual observation fixed only on color so deceiving, however, is that silver is, perhaps, the most dissimilar metal to tin. For silver will alloy with every metal with little change in their natures—according to Biringuccio. Tin, on the other hand, will embrittle the other metals ("... acts like a powerful poison which envenoms and corrupts the others...").

Referencing unnamed "Speculators," he explained tin's ability to embrittle the other metals using, apparently, the Aristotelian notion of Water. (This argues against the conjecture about Paracelsus's using the Aristotelian notion of Fire in the description of Tin.)

> *"... this fact (Tin's embrittling effects) arises from its great wateriness, which is very subtle and poorly decocted and almost like that of quicksilver. This wateriness, then, by means of its subtlety which joins itself with metals, imposes itself into that unctuous and viscous material which renders them flexible, and thus deprives them of their toughness and corrupts them, excepting Lead, in such a way that it almost converts them into another nature. Although Tin alters Lead it does not apparently act on it so much, since both have almost an equal and proportioned similarity of nature. For this reason Tin is called "white lead" by the alchemists.*[56]

To avoid the impression that tin embrittled all other metals, Biringuccio pointed out the exception of lead. Even though tin does act on lead, it does not appreciably change lead, since both tin and lead share similar natures. For this reason, he explained, tin was known as "White Lead."

Continuing his description of tin, Biringuccio commented on tin's malleability and the ease by which it is worked:

> *... it is easily worked since it melts with any kind of fire, and with little effort. This metal, either pure or when mixed with lead, stands up well under the hammer, so that, if desired, it can be spread out thinner than paper.*[57]

This corresponds well with Paracelsus's description of the "fragility" of tin. Since tin is so ductile, it is easily worked. Further, because of its relatively low melting temperature, any heating increases the ease by which it can be shaped and yet stand up to the imposed stresses from working, allowing it to be worked into thin leaves or sheets.

Further on in *Pirotechnia,* Biringuccio expounded on additional curious effects tin has on other metals: making them sonorous and "whitening" them (i.e., imparting a silvery color):

> *... it produces no sound itself ... but by hardening other metals it makes them sonorous, just as if it puts the spirit there and vitalizes the substances. Thus by the mixture of two bodies there comes the creation of a third that is neither one nor the other, indeed it is entirely different, brittle and much harder than either one of them was before.*
>
> *The whiteness that tin introduces into metals results from its spreading out in this case as a watery or very subtle thing, overcoming the redness of the copper or the yellowness of the gold, and extinguishing it, making the metal very white, from red or yellow, in such a way that it shows even more whiteness than the tin itself previously showed.*[58]

Biringuccio stated here that tin could emit its characteristic creaking sound on its own but could also impart this sound to other metals. This is curious, since, later in *Pirotechnia* he did affirm tin's unique crackling sound: "*... when it is bent or some thin end of it is held tightly by the teeth, is heard to crackle as frozen water does ... it (is) a sign for recognizing when tin is pure or mixed.*"[59]

The "whitening" involved the coating of metals such as silver and gold with molten tin that would spread out over the surface as water would. Biringuccio believed that the whiter appearance was because of the greater shininess and brightness that could be achieved from the heightened polish that the greater hardness could receive.[60]

CHAPTER 5

Martinus Rulandus (German Physician and Alchemist, 1532-1602)
Rulandus did not offer an explanation for "Tin's cry," but he did provide terms and definitions associated with tin. According to Rulandus, tin differed from White Lead and was a white metallic substance, impure and livid. Also, tin was the product "... *of pure Quicksilver (fixed or not), clear and white, and of a Red Sulphur, of slight fixity, not pure, and unable to prevail over the Quicksilver.* ..." Here, Rulandus was defining tin in terms of the principles of Mercury and Sulphur with Mercury predominating. This would account for the metallic nature of tin. Rulandus interchanged the Latin term *Stannum* for tin with references to lead—"Stannum Dives" (i.e., Rich Lead); Stannum mediocre for Raw, Workable Lead; and Stannum pauper for Uncalcined Lead. In regard to the source of tin, Rulandus referenced Serapion who contrasted the mining of Stannum from the earth with White lead, which is not mined but produced from Galena, Pyrites, etc.[61]

Nicholas Lemery (French Chemist, 17 November 1645 – 19 June 1715)
One of the most complete discussions of tin in later alchemical texts was that found in Nicolas Lemery's *Course of Chymistry*. Chapter Three of this work dealt specifically with tin. Lemery began his description by recalling the name accorded to it by the early alchemists—White Lead. According to Lemery, tin is a metal that is closest to silver in color but differs significantly from silver in the "... Figure of its Pores and in the Solidity and Weight." (In Lemery's thought, one of the characteristics of substances was the configuration of their pores. Substances had pores of various sizes and shapes that would determine the extent to which they could be penetrated and dissolved. Gold, for instance, had very restricted pores thus rendering it corrosion-resistant and impervious to most acids.) He also reiterated tin's traditional sympathetic connection with the planet Jupiter, from which, it was traditionally believed, tin received its particular influence. Further, tin was considered to be a malleable substance, sulphurous, and easily put into fusion (melted).

In addition, his description included the types of tin—Plate Tin, Common Tin, and Bell-Metal. The best tin, according to Lemery, was the kind that came from Cornwall in England. It was called Plate Tin and was to be preferred for chemical operations. Common Tin contained small amounts of lead and was alloyed with yellow brass. Bell-Metal was tin alloyed with bismuth, antimony, or some other metallic matter (Tin was particularly useful in this application since it rendered the bell-metal hard, solid, and compact, which allowed the metal to "... yield a Sound ..." when struck).

The chapter also described various operations and reaction products involving tin, including pulverization, calcination, sublimation, Salt of Jupiter, the Magistry of Jupiter, Flowers of Jupiter, Oil of Tin, and Jupiter Diaphoretick. Before tin could be used in subsequent operations, it had to be properly prepared by pulverization, calcination, or sublimation.

Under the section, "Pulverization of Tin," Lemery explained that tin, being malleable, could not be reduced to a powder by the "... usual ways of powdering...." Thus, he had to provide a recipe by which tin could be reduced to a powder:

> *Melt in a Crucible what Quantity of Tin you think fit, and cast it into a round wooden Box, that has been rubbed within on all Sides with a Piece of Chalk, enough to whiten it, cover this Box, and presently shake it about, until your Tin is become cold, and so you'll find it converted into a gray powder.*[62]

In the presence of carbonate ions (CO_3^{2-})—from the chalk—both tin (II) and tin (IV) react differently than would be expected for such metallic ions such as Pb^{2+}. The chalk (calcium carbonate) would release carbon dioxide upon heating (a thermal decomposition reaction, or calcination, to above 840°C in the case of $CaCO_3$), to form calcium oxide, CaO. Since tin melts at 232°C, the mixture with tin would allow any unstable tin carbonate to also decompose to tin oxide and carbon dioxide.

Thus, in this case, the reaction products would be carbon dioxide and the corresponding oxide. The gray-colored powder Lemery observed presumably was tin (IV) oxide since its color can be yellow or light gray. Tin (II) would not be a likely candidate since it is black or red in the anhydrous state or white in the hydrated state.[63]

In his remarks on the calcination of tin, Lemery noted that the calcined tin gained weight in spite of his observation that "... the Fire evaporate(d) a part of its Sulphur...." He reasoned that in place of the sulphur "... a greater Quantity of igneous Particles..." were acquired by the tin. In another experiment involving the calcination of powdered tin mixed with an equal amount of dissolved silver, precipitated with salt water and dried, but retaining some moisture, the mixture became so hot it ignited on its own. Using his theory of acids, in which the acid was imagined to be like a cog with spikes, he explained this phenomenon as caused by "... the remaining Points of the Spirit of Nitre and Sea salt, which are incorporated with the precipitate Silver, and which fixing upon the Tin, do penetrate its Pores with such Violence, that they kindle the Sulphur."[64]

According to Lemery, the sublimation of tin was achieved by means of a volatile salt, which in his recipe was sal armoniac (ammonium chloride, NH_4Cl):

> *Take one part of tin, and two parts of sal armoniack in powder: mix them well together, and put your mixture into a strong earthen cucurbit, that is able to endure the fire, and whose two-thirds at least do remain empty, fit unto it a blind head, lute the conjunctions exactly well, and place your vessel on the grate in a small furnace with open fire . . . till the bottom of the cucurbit is grown red hot, and continue such a fire till nothing more will sublime, which you'll know by the head's growing cool, and then the sublimation is at an end. Let the vessels cool, and so unlute them, you'll find flowers stuck to the head, and to the top of the body, that are nothing else but some parts of tin raised up by the sal armoniack, and at the bottom of the body you'll find some tin revived.*[65]

Lemery's "sal armoniack" was camel urine.[66] Assuming urine is fairly standard across mammals, the primary constituent besides water would be urea (H_2NCONH_2 or $CO(NH_2)_2$). The "flowers" described the result of the reaction, which according to Lemery's use of that term elsewhere, would have been "flowers of Jupiter" or tin (IV) oxide.

The Salt of Jupiter was the product of tin exposed to acids and reduced into the form of a salt. For Salt of Jupiter, Lemery recommended using only Cornwall or English Tin since the common tin would ". . . give only a Green and Sharp Salt, because it contain(ed) some Quantity of Copper." To produce this salt, the calx of tin was dissolved in vinegar. The calx, however, had to be thoroughly calcined so that, according to Lemery, some quantity of its sulphur was lost. Otherwise, the "Acidity of Vinegar" would not have been able to dissolve the calcined tin. If the dissolved tin were allowed to dry, crystals would form, which were most likely tin acetate ($Sn(CH_3COO)_4$).

If the vinegar-treated calx of tin was exposed to the Oil of Tartar (a saturated solution of potassium carbonate, K_2CO_3) per deliquium (liquefaction through absorption of moisture from the air), the Magistry of Jupiter would result. According to Lemery, the Oil of Tartar, which is an alkali, destroyed the "Acid of Vinegar" which kept the tin dissolved, and ". . . forc(ed) it to let go its hold...." Thus the operation involved tin dissolved by an acid and precipitated by an alkali salt. Lemery offered the following recipe:

> *Dissolve the Flowers of Tin in a sufficient quantity of Water, Filtrate the Dissolution, and pour upon it drop by drop the Spirit of Sal Ammoniac, or the Oil of Tartar made per Deliquium, there will precipitate a very white Powder. You must Edulcorate[67] it by washing it several times with warm Water, and afterwards dry it. It serves for Paint...[68]*

This recipe begins with the Flowers of Tin, which has already been generated by an operation that volatized tin by means of a volatile salt. Lemery offered the following recipe for the production of Flowers of Jupiter or tin:

> *... mix a Pound of Tin and two Pounds of purified Salt-peter, throw a Spoonful of the Mixture through the hole of the Pot, and stop it, a Detonation soon follows, which when it is over, throw in another Spoonful, so as to continue to do, until all the Mixture be spent, let the Vessels cool, and unlute them, and you'll find in the Receiver a little Spirit of Nitre,[69] and in and around about the Alludel, a very white Flower of Tin, gather them together with a Feather, then wash them diverse Times with Fountain Water, and when you have dried them on Paper in the Shade, keep them in a Viol, they serve for Paint...[70]*

The recipe first required purified salt-peter, which would have been pure potassium nitrate. This could have been prepared by boiling the salt-peter in a small amount of water and then reacting it with potassium carbonate from wood ashes. This would remove calcium and magnesium salts[71] as precipitates, leaving a purified potassium nitrate. Potassium nitrate is easily ignitable, explaining the immediate detonation Lemery observed. The assumption is that the potassium nitrate decomposes to potassium nitride (KNO_2) and oxygen (O_2) in the vessel, leaving the tin to combine with the oxygen to form white tin (IV) oxide[72], also known as Flowers of Tin.[73]

The Oil of Tin, also known as the Liquor of Tin, was prepared by treating tin with *Aqua regia*, a mixture of hydrochloric (Spirit of Salt) and nitric (*Aqua fortis*) acids:

> *This Operation is Tin reduced into a thick Fluid by the Means of Aqua Regia.*
>
> *Put into a Glass Vessel what quantity you please of plated Tin, cut into small Pieces, pour upon it thrice as much of Aqua Regia, made of two Parts of Aqua Fortis, and one Part of Spirit of Salt. Place the Vessel upon a small Fire of Digestion and there will be a slow Ebullition, and the Tin will dissolve by Degrees... evaporate the Humidity in a Sand Heat, and there will remain a whitish gray Salt...*[74]

In his remarks on this recipe, Lemery applied his theory of acids. Since tin is sulphurous and soft, he explained, it softens the "points" of the *Aqua regia* corpuscles, causing the reaction to proceed slowly. He went on to observe that the dissolution of tin was like that of antimony, for both precipitate a white powder (in the case of the recipe above this was tin (IV) oxide) in the bottom of the vessel. Further, drying the precipitate reduced the oxide to the consistency of salt. But if moisture were added later, the moistened oxide would have the consistency of oil or liquor, which Lemery averred were improper since the oil of tin was only tin dissolved by acid spirits.[75] The oil of tin, Lemery claimed, was good against the "Rottenness of the Teeth" and "... for cleaning away and eating proud Flesh."[76]

The last tin compound Lemery addressed in his discussion on tin was Jupiter Diaphoretick, which was a mixture of tin and the Regulus of Antimony with Mars fixed by salt-peter:

> *Take the finest tin, and of the Regulus of Antimony with Mars... of each eight ounces, dissolve them together in a crucible upon a charcoal-fire, and pour the matter dissolved into an iron mortar, heated and greased, let it cool, then beat it into a powder, and mix with it three times as much salt-peter purified....*[77]

The Regulus of Antimony with Mars was an antimony-iron compound.[78] According to Lemery, the result was useful for treating diseases of the liver, malignant fevers, and smallpox.

Conclusion

The alchemists believed tin was similar to silver, but was in an impure state that could be transmuted into silver. Many of them thought that because the nature of silver was close to the nature of gold, tin could be, eventually, through mediatorial operations, transmuted into gold. In pursuit of that goal, they developed various recipes that produced what they believed was at least an intermediate state—silver. Though the actual result was only a silvery coating or coloration of the metal being treated, other significant uses were discovered. For instance, with the use of mercury, the transmutational alchemists came up with amalgams of silver and tin—similar to dental amalgams. For the physician-alchemist, medical treatments using tin compounds were uncovered—treating diseases of the liver, malignant fevers, smallpox, and removal of persistent granulated scabs.

In addition, one other pursuit deserves appreciation, even though it did not reach a satisfactory result—an explanation of the "Tin cry." Of course, it was not possible for the alchemist to uncover the cause of this mysterious sound that tin would make when bent or deformed. The necessary equipment, theories, and methods were not available that have allowed us, today, to understand the nature of twinning. Yet, credit is due to the alchemists for their attempts to explain the phenomenon. Surprisingly, one alchemist (Avicenna), who must have had incredible intuition, came close to understanding a part of the mechanism involved. We know now that other metals also share this characteristic. But in antiquity only tin was known to "cry"—the others did not, unless they were alloyed with tin or treated in some other manner. Though the endeavors to uncover Jupiter's mien did not always meet with success in the hands of the alchemists, they were foundational, providing building blocks upon which further knowledge and theories could be built.

CHAPTER 5

Notes

1. The word "tin" comes from the Anglo-Saxon "tin," and the chemical symbol Sn stems from the Latin *Stannum*, which may have derived from the Sanskrit word "stan," meaning hard, or from the Cornish word "stean," which was known to the Romans.
2. Valera, R.G.; Valera, P.G. (2003), "Tin in the Mediterranean area: history and geology", in Giumlia-Mair, A.; Lo Schiavo, F. (eds.), *The Problem of Early Tin*, Oxford: Archaeopress, 3–14.
3. Muhly, J.D. (1985), "Sources of tin and the beginnings of bronze metallurgy", *American Journal of Archaeology*, vol. 89, no. 2, 275–291. Also, Wertime, T.A. (1979), "The search for ancient tin: the geographic and historic boundaries", in Franklin, A.D.; Olin, J.S.; Wertime, T.A. (eds.), The Search for Ancient Tin, Washington, D.C.: A seminar organized by Theodore A. Wertime and held at the Smithsonian Institution and the National Bureau of Standards, Washington D.C. March 14–15, 1977, 14–15.
4. John Emsley, *Nature's Building Blocks: An A-Z Guide to the Elements*, (Oxford: Oxford University Press, 2011), 552.
5. Thomas Courtney, *Mechanical Behavior of Materials*, 2nd ed. (McGraw-Hill, 2000), See also, Donald R. Askeland and Wendelin J. Wright, *The Science and Engineering of Materials*, Seventh Edition, (Boston: Cengage Learning, 2016), 127.
6. Donald R. Askeland and Wendelin J. Wright, 127.
7. Cathy Cobb, Monty L. Fetterolf, and Harold Goldwhite, *The Chemistry of Alchemy: From Dragon's Blood to Donkey Dung: How Chemistry was Forged,* (Amherst, NY: Prometheus Books, 2014), 51, 205-206.
8. Pliny, *Natural History*, Book IV, 104, 112, 119.
9. Pliny, *Natural History*, Book XXXIV, 47, 157.
10. Pliny, *Natural History*, Book XXXIV, 97.
11. Philip Wheeler, *Alchemical Symbols*, Fourth Edition, R.A.M.S. Library of Alchemy, Vol. 21, (Kansas City, MO: R.A.M.S. Publishing Co., 2018), 79.
12. Matteo Martelli, *The Four Books of Pseudo-Democritus, Sources of Alchemy and Chemistry, Ambix,* Vol. 60, Supplement 1, 2013, S105-S107.
13. Matteo Martelli (2009) "'Divine Water' in the Alchemical Writings of Pseudo-Democritus," *Ambix*, 56:1, 5-22.
14. Olivier Dufault (2015) "Transmutation Theory in the Greek Alchemical Corpus," *Ambix*, 62:3, 215-244.

15. Wheeler, 79.
16. Raphael Patai, *The Jewish Alchemists: A History and Source Book*, (Princeton, NJ: Princeton University Press, 1994), 125, 133-134.
17. Marianne Marchini, et al. "Exploring the Ancient Chemistry of Mercury," *PNAS*, Vol. 119, No. 24. www.pnas.org. Retrieved 2-Oct-2023.
18. Zosimus classified different types of mercury in accordance with the metals used in its extraction. In a long list of code names used for mercury, he also refers to it as "water of copper" and "water of lead", and he mentions various names given to "mercury from copper", "mercury from lead". Marianne Marchini, et al.
19. H. E. Stapleton, R. F. Azo, M. Hidāyat Husain, & G. L. Lewis (1962) "Two Alchemical Treatises Attributed to Avicenna," *Ambix*, 10:2, 41-82, 46n1.
20. Ibid., 48n4.
21. In her translation, Gail Marlow Taylor identified sal-ammoniac with alum. 126, 128.
22. There seems to be disagreement on the meaning and use of yellow sulfur among translations. Gail Marlow Taylor's translation of this section identifies yellow arsenic sulfide to be used for "whitening" and red arsenic sulfide for "reddening." See Gail Marlow Taylor, *The Alchemy of Al-Razi: A Translation of the "Book of Secrets,"* (North Charleston, North Carolina: Createspace Independent Publishing Platform, 2014), 126.
23. H. E. Stapleton, R. F. Azo, M. Hidāyat Husain, & G. L. Lewis, 59n30. Also, Trituration—the production of a homogeneous powdered material by mixing and grinding component materials thoroughly.
24. Gail Marlow Taylor, 167.
25. Wheeler.
26. Isaac Asimov, *Asimov's Biographical Encyclopedia of Science & Technology (The Lives & Achievements of 1510 Great Scientists from Ancient Times to the Present)*, Second Revised Edition, (Garden City, New York: Doubleday & Co., Inc., 1982), 53. See also, Nasr, Seyyed Hossein (2007). "Avicenna". Encyclopedia Britannica Online. Retrieved 5-Oct-2023.
27. H. E. Stapleton, R. F. Azo, M. Hidāyat Husain, & G. L. Lewis (1962) Two Alchemical Treatises Attributed to Avicenna, *Ambix*, 10:2, 41-82, 46n1.
28. Ibid, 47-48.

29. Ibid, 46n1.
30. John Eric Holmyard, *Makers of Chemistry*, (Oxford: Clarendon Press, 1931), 73-74.
31. Raphael Patai, 98.
32. Ibid, 106.
33. Ibid, 107.
34. Wheeler.
35. E.J. Holmyard and Richard Russell, *The Works of Geber*, (1928) Kessinger Legacy Reprints, (Kessinger Publishing), 66-67.
36. Ibid, 138-139.
37. William R. Newman and Lawrence M. Principe, *Alchemy Tried in the Fire: Starkey, Boyle, and the Fate of Helmontian Chymistry*, (Chicago: The University of Chicago Press, 2002), 44-45. See also, William R. Newman, *Atoms and Alchemy: Chymistry and the Experimental Origins of the Scientific Revolution*, (Chicago: The University of Chicago Press, 2006), 33.
38. E.J. Holmyard and Richard Russell, 140.
39. William R. Newman and Lawrence M. Principe, 46.
40. Raphael Patai, 15-16.
41. Ibid., 309.
42. Wheeler.
43. Paracelsus, *The Hermetic and Alchemical Writings of Aureolis Philippus Theophrastus Bombast, of Hohenheim, called Paracelsus the Great*, in two volumes, edited by Arthur Edward Waite, (Mansfield Centre, CT: Martino Publishing, 2009), Vol. I, 78-79.
44. John Emsley, *Nature's Building Blocks: An A-Z Guide to the Elements*, 553-554.
45. John Emsley, *The Elements of Murder: A History of Poison*, (Oxford: Oxford University Press, 2005), 384.
46. Paracelsus, Vol. II, 15.
47. Vannoccio Biringuccio, *Pirotechnia*, edited by Cyril Stanley Smith and Martha Teach Gnudi (Cambridge, Massachusetts: MIT Press, 1959), 186-187. ". . . if you wish this acid to be good and to work well, it is necessary to add half a penny weight of silver to every pound of acid. In order to do this, take one or two pounds or however much you wish of this acid in a small cucurbit and add to it the entire weight of silver (granulated or beaten with a hammer) that belongs to the whole

quantity of acid that you have made. As soon as this is in it, you will see the acid begin to grow turbid and to make its power felt. Even if you left it alone, it would affect its operation, but it will do it more quickly and better if you put it over hot ashes. A short time after you have put it there, you will see all the silver dissolve into water. After it is dissolved and the acid has been left alone, you will see a grossness similar to a very white chalk fall to the bottom. When this has fallen down and the acid made clear with this substance of silver, decant it very slowly into the receiver where the whole quantity of aqua fortis is. You will see all this alter just as happened with the small amount. After it has stood a short time, you will see a coarse precipitate of a very white material like the aforesaid go to the bottom."

48. Rebounding or reflecting heat or flame onto a metal. Also, to heat, melt, or refine (a metal or ore) in a reverberatory furnace.
49. T. P. Sherlock (1948) "The Chemical Work of Paracelsus," *Ambix*, 3:1-2, 33-63.
50. Paracelsus, Vol. II, 15-16.
51. Sherlock.
52. Gold (III) Chloride Hydrate, Stanford Advanced Materials, https://samaterials.com/gold-chloride-hydrate-powder.html. Retrieved 29-Sep-2023. See also Gold (III) Chloride, Wikipedia, Retrieved 29-Sep-2023.
53. Common Chemistry, https://commonchemistry.cas.org. Retrieved 29-Sep-2023.
54. Sherlock.
55. Vannoccio Biringuccio, *Pirotechnia*, Edited by Cyril Stanley Smith and Martha Teach Gnudi, (Cambridge, Massachusetts: Cambridge University Press, 1966), 59.
56. Ibid, 59-60.
57. Ibid.
58. Ibid, 60-61.
59. Ibid, 211.
60. Ibid, 61.
61. Martinus Rulandus, *A Lexicon of Alchemy, Containing a Full and Plain Explanation of All Obscure Words, Hermetic Subjects, and Arcane Phrases of Paracelsus*, (Zachariah Palthenus, bookseller, in the free republic of Frankfurt, 1612), 317.

CHAPTER 5

62. Nicolas Lemery, *A Course of Chymistry, The Fourth Edition*, (London, 1720), Gale Eighteenth Century Collection. 82.
63. "Characteristic Reactions of Tin Ions," and "Occurrence, Preparation, and Properties of Carbonates," Libretexts. https://chem.libretexts.org/Bookshelves/General_Chemistry/Chemistry. Retrieved 21-Sep-2023.
64. Lemery, 81-84.
65. Ibid, 85.
66. Ibid, 274.
67. Edulcorate: To make more palatable or acceptable. A chemical operation in which salty or sour materials are removed from a product to leave a "sweetened" (generally meaning tasteless in this context) substance. Edulcoration may be carried out by simple washing with water, by repeated distillations of water or spirit of wine, or by other means. The Chymistry of Isaac Newton. https://webapp1.dlib.indiana.edu/newton/reference/glossary.do. Retrieved 4 Sept 2023.
68. Lemery, 85.
69. Nitric acid, HNO_3.
70. Lemery, 86.
71. Salt-Peter can be composed of potassium, calcium, and/or magnesium nitrates.
72. Tin (IV) oxide may also appear yellow or light-gray in color.
73. CAMEO: Conservation and Art Material Encyclopedia Online, Museum of Fine Arts, Boston. Description. White powder that is often incorrectly called tin oxide. Stannic oxide, or tin dioxide, occurs in nature as the mineral cassiterite. It is used as an abrasive, sometimes in mixtures with lead oxide, for polishing glass, marble, silver, and jewelry. Stannic oxide is also used as a mordant for dyeing fabrics and as a weighting agent. Additionally, it is used as an opacifier in glass and glazes to produce a translucent milky color. Tin dioxide reacts with chrome oxides to produce a ruby red color in glass and glazes. Synonyms and Related Terms--stannic anhydride; tin peroxide; tin dioxide; tin oxide (sp); tin (IV) oxide; stannic acid; putty powder; putty; jeweler's putty; white tin oxide; flowers of tin; polishing powder; tin https://web.archive.org/web/20121104011622/http://cameo.mfa.org/browse/record.asp?subkey=8907. Retrieved 20-Sep-2023.
74. Lemery, 87.

75. Ibid.
76. Randy Jacobs, MD, Dermatology, "Proud flesh is also known as persistent granulation tissue, and occurs when the scabs normal granulation tissue does not go away." https://www.randyjacobsmd.com/granulation-tissue.htm. Retrieved 22-Sep-2023.
77. Lemery, 88.
78. Ibid, 181-182.

Chapter 6

Iron, Vulgar, Yet a Great Metal

Introduction

Mars, the Roman god of war, arose to that exalted position from humble origins. From the mists of myth, he initially emerged as an agricultural deity, protecting the crops and ensuring the prosperity of harvests for a pastoral culture. As protector, the people turned to him for courage and strength to face and fight those who sought to invade their lands and to be their conquerors. They saw the spirit of Mars in their courage, their strength, and their weapons and tools. Mars was looked to, then, as a power for securing the peace, for would-be invaders would fear the people's resolve and weapons.[1] It was fitting that Mars came to be associated with iron, the metal of choice for weapons, tools, and implements, being the strongest and hardest of the seven metals known in antiquity.

Mars[2]

Iron has been known and utilized since at least 3500 BCE throughout the ancient world. Egyptian artifacts have been discovered dating back to that time. The Egyptians called iron "the metal of heaven," which indicated that one source of their iron was from meteors (though they would not have known meteors as such, they did make the observation that "rocks fell from heaven." This has been substantiated by the presence of nickel alloyed with iron in archeological artifacts).[3] The Hittites of Asia Minor may have been the first to smelt iron from its ores around 1500 BCE.[4]

Compared to the other known metals, iron was an anomaly. It had a much higher melting point and much greater strength than the other known metals (gold, silver, mercury, copper, tin, and lead). These qualities naturally drove the alchemists to seek explanations. Because they believed that each of the metals had a peculiar "spirit" that could be beneficial and transferred to other substances, many of the alchemists sought medical uses for iron, much like they did for the other metals. The following discussion is a survey of the theories of iron's formation and nature, as well as medicinal substances discovered along the way.

Abufalah (11th century CE)

Though ancient and medieval alchemists offered various explanations for the origin and nature of iron, the principles of Mercury and Sulfur (Mercury was the volatile and metallic constituent of bodies, carried the quality of liquidity, while Sulfur was the combustible nature; it denoted expansive force, evaporation, and dissolution. See the chapters "The Formation of the Metals: An Alchemical Perspective" and "Tria Prima.") were common to many of them. These principles were believed to have originated with Jabir ibn Hayyan, a Persian polymath and alchemist of the 9th century CE (it is unclear as to whether he was a historical personage). Whether Jabir actually existed or not, the principles of Mercury and Sulfur were known by at least the 11th century.

The Arabic alchemist, Abufalah, theorized that all metals were formed from quicksilver (mercury) and sulfur. He did not originate the theory himself, he explained; he attributed it to "some sages." Then he expounded on it at great length. In addition to the principles of Mercury and Sulfur, Abufalah relied on the Aristotelian elements—Earth, Fire, Air, and Water—and the qualities of Dry, Wet, Cold, and Hot. In effect, his theory on the origins of the metals became an addendum to his overall theory on the origin of all things. He began with the assertion that all things–plants and minerals–were controlled by the power of the zodiac exerted through the agents of the elements.

The first things to come into being were two kinds of vapors—the moist and the dry. The moist vapors, according to Abufalah, rose from the earth to produce rain, and the dry vapors rose from the earth to produce thunder and wind. When the moist and dry vapors came into being in the earth, Abufalah explained that the dry and smoky vapor produced the mineral stones such as red arsenic and marcasite. The moist vapor, on the other hand, produced two kinds of metals. One kind melted like copper, gold, and silver. The other "broke like iron" (perhaps a reference to the lack of malleability of iron compared to the other metals). In regard to the metals' appearance, Abufalah saw the direct influence of the "moving stars" (the planets and the Moon), so the appearance of gold was due to the "Sun and its sparks," the whiteness of silver was due to the Moon, the blackness of lead was due to Saturn, the redness of iron (iron oxide) was due to Mars, the greenness of tin was due to Jupiter, and the blueness of copper was due to Venus.

The metals, then, were formed by the "moist vapors that develop in the space of the Earth, (which) are affected by the heat of origin ... (that) ... ripens them and purifies them during the long time of their stay there...." In this manner, the metals took on weight, thickness, and purity of appearance. In regard to iron, Abufalah explained its formation as a result of the quicksilver and sulfur being exposed to cold before they mixed prior to ripening.

Abufalah explained the differences in the melting points and density of the metals by using the concepts of dryness, moistness, airiness, and attachment. For instance, the reason gold, silver, and iron took longer to melt than tin or lead was due to the degree of attachment of the dry "dusty parts" to the "watery and airy parts." In gold, silver, and iron, the dusty parts were well mixed, better attached, and better fused to the "watery and airy parts," than they were in tin and lead. Likewise, Abufalah explained the differences in heaviness (density) by the degree of attachment of the parts and the abundance of the "dusty parts." The heavier (more dense) metals had stronger attachments of their parts and more "dusty parts." This was why, in Abufalah's thinking, gold did not deteriorate, lead was so heavy, and iron was so strong.[6]

Pseudo-Geber (13th century CE)

Compared to Abufalah, Pseudo-Geber offered a variation of the Mercury-Sulfur theory for the formation of iron. When the fixed (non-volatile) earthy Sulfur mixed with fixed earthy Mercury (Argent vive), and both of these were not pure but of a livid[7] "Whiteness," and the Sulfur was highly fixed, then–according to Pseudo-Geber–iron was the result. The high fixity of Sulfur prohibited fusion, which explained the high melting point of iron and caused it to be the least favored of the metals for transmutation.[8] In other words, he considered iron the most difficult of the metals:

> "... it [Mars or Iron] is a Metallic Body, very livid, a little red, partaking of Whiteness, not pure, sustaining Ignition, fusible with no right fusion, under the Hammer extensible, and sounding much. But Mars is hard to be handled, by reason of the Impotency of its fusion [i.e., high melting temp.], which if it be made to flow by a Medicine changing its Nature, is conjoined to Sol and Luna, and not separated by Examen, without great Industry ...
>
> Yet Mars, among all Bodies, is of least Perfection in Transmutation, to be handled most difficult, and of exceeding long Labor. Therefore, whatsoever, Bodies are more remote from swiftness of Liquefaction [ease of melting], they are found of more difficult handling in the Work of Transmutation.[9]

In contrast to Abufalah, however, Pseudo-Geber explained that the difference in the melting points of the metals was due to the abundance of Sulfur present. Since Sulfur, by "the work of Fixation," (rendering a substance less volatile) impedes or enhances fusion (melting), the more there is of fixed Sulfur, the less volatile the substance becomes. Pseudo-Geber quickly pointed out, however, that Sulfur was not fixed (i.e., it burns) and thus by fixed Sulfur he meant Sulfur that has been calcined (Here, is an instance in which the principle of Sulfur and the element of sulfur have been either confused or conflated):

> Therefore, hence 'tis manifest that Sulphur, by the work of Fixation, more swiftly destroys the easiness of Liquefaction than Argentvive. By these is manifested the Cause of Swiftness and Slowness of Fusion in every Body. For what hath more of fixed Sulphur, more slowly admits of Fusion, than what partakes of burning Sulphur, which more easily and sooner flows: and this is clearly enough already declared by Us. But that the fixed Sulphur makes slower Fusion is evident by this, viz. that it is never fixed unless it be calcined, and no Calcinate gives Fusion: therefore in all Things it must be impeded the same.[10]

Though Pseudo-Geber considered iron not amenable to transmutation, he asserted, nevertheless, that the imperfect metals (lead, tin, iron, and copper) could be altered and perfected by their own particular "Medicine" that would address the deficiency within each of the metals. Of course, his hope was to find the one "Medicine" by which all metals could be transmuted. First, the metal had to be prepared, however. He outlined three methods for the preparation of iron. The first involved sublimation with arsenic (most likely an arsenic sulfide), the second involved multiple sublimations with arsenic until a quantity of the arsenic remained, and the third involved fusion (melting) with lead and tutia (copper sulfate). Softening the iron was the first step ("Alterations of the first Order") toward the Magistry (the philosopher's stone or elixir). To accomplish this, the iron had to "... be conjoined and sublimed often with Arsenik."[11] Presumably, the iron (He doesn't say whether he is using iron filings or an iron ore) reduced the arsenic sulfide, resulting in metallic arsenic and iron sulfide.

Paracelsus (c. 1493 – 24 September 1541)
With Paracelsus, another substance, in addition to Mercury and Sulfur, was introduced to explain the nature of iron—Salt (Principle of Fixity). According to Paracelsus, iron consisted of a larger proportion of Salt and Mercury than of Sulfur. The order of the constituents of iron, he maintained, was "its own body," which preponderates, followed by Salt, then Mercury, and last, Sulfur. He pointed out, however, that the extent of the Salt was circumscribed by the Sulfur. If there was more Salt than the composition of the Sulfur required, then the metal could not be formed. He reasoned that the flexibility (malleability) of the metal depended on its Mercury and its coagulation on Salt. Too much Salt and the metal became too hard and brittle:

> *In the generation of Iron, there is a larger proportion of Salt and Mercury, while the red Sulphur from which Copper proceeds is present in a smaller quantity. It contains a cuprine salt, but not in equal proportion with Mercury. Its constituents are its own body, which preponderates; then comes Salt, afterwards Mercury, and, lastly, Sulphur. When there is more Salt than the composition of Sulphur requires, the metal can in no wise be made, for it depends upon an equal weight of each. For flexibility proceeds from Mercury and coagulation from Salt. Accordingly, if there be too much Salt, it becomes too hard.*[12]

In regard to the observed properties of iron, Paracelsus gave a rather obscure account. In the description from *The Coelum Philosophorum*, "The Third Canon: Concerning Mars and His Properties," he remarked that iron, compared to the other metals, had little "efficacy." Did he mean, like Pseudo-Geber, that iron was the least amenable to transmutation compared to the other metals? But from the standpoint of iron's uses, iron could arguably be seen as more efficacious than the six other metals.

He also asserted, in an occult (i.e., hidden) and mystical way, iron derived its weight, its degree of solidity ("... strength of coagulation..."), and its degree of hardness from the other metals that had yielded a degree of their own solidity and hardness to iron.

> *The six occult metals have expelled the seventh from them, and have made it corporeal, leaving it little efficacy, and imposing on it great hardness and weight. This being the case, they have shaken off all their own strength of coagulation and hardness, which they manifest in this other body. On the contrary, they have retained in themselves their color and liquefaction, together with their nobility.... Mars acquires dominion with a strong and pugnacious hand, and seizes on the position of king. He should, however, be on his guard against snares, that he be not led captive suddenly and unexpectedly.*[13]

Presumably, Paracelsus was alluding to iron having a relative preponderance of salt (the principle of coagulation and solidity), compared to the other metals that yielded a portion of their salt. His *De morbis metalicis* may shed some light on the above statement. From that work, he purports that all metals can be generated from mercury by the application of heat.[14] If we understand that all metals are latent within mercury, then as they congeal out of the mercury, each metal would have to give up, in varying degrees, certain qualities to the other metals to explain the differences between the congealed metals. For instance, every metal has the property of solidity (coagulation) but differs in their melting points, which would be indicative of their relative "... strength of coagulation...." Iron has the highest melting point of the seven metals (mercury, tin, lead, copper, silver, gold, iron) and the greatest hardness, thus Paracelsus believed his explanation accounted for the observed properties of iron. The metals retained, however, their own appearances and ease of liquefaction. In spite of the negative remark concerning iron's efficacy, Paracelsus affirmed that iron had acquired dominion and "the position of king." Here, perhaps, he had acknowledged the role of iron in everyday life. But the alchemist, he warned, had to be careful. Iron easily decomposes, reacting with oxygen to form iron oxide, and thus loses its strength.

CHAPTER 6

Though Paracelsus was interested in the transmutation of metals, his greater concern was iatrochymistry, the utilization of substances for medical treatments. He added the metal with "little efficacy" to his array of medicinal treatments. The Crocus of Mars (iron oxide), he explained, had to be prepared first. He provided three methods for doing this. The first involved simply heating thin plates of steel until they became red-hot and then extinguishing them in vinegar made from wine. When the vinegar became red, he distilled it and then coagulated the residue into a dry powder. The redness would have indicated that the iron had oxidized to iron (III) oxide (Fe_2O_3):

> ... it should be known concerning Iron that it can be mortified and reduced to a crocus in the following way: Form very thin plates of steel, beat them red-hot, and then extinguish them in vinegar made from wine. Keep doing this until you see the vinegar has become very red. When you have enough of this red vinegar, pour it all out, and distill therefrom the moisture of the vinegar. Coagulate the residuum into a dry powder.[15]

Crocus Martini

Paracelsus's second method for preparing the Crocus of Mars involved coating steel plates with equal amounts of sulfur and tartar (most likely potassium bitartrate, $KC_4H_5O_6$, also known as wine-stone, a deposit that forms inside wine barrels), and then reverberating the coated plates:

> There is, however, another way of making the Crocus of Mars which partly surpasses the former, and is carried out with much less expense and labor, thus: Stratify very thin plates of steel with equal quantities of sulfur and tartar. Afterwards, reverberate. This produces the most beautiful crocus, which should be taken from the plates.[16]

The final method was to smear a plate of iron or steel with *aqua fortis* (nitric acid). This method would have resulted in iron (III) nitrate, which is yellow in color. The nitrate could then be easily removed or decomposed over time. When heated to near the boiling point, the nitric acid would evaporate and leave a residue of iron (III) oxide (Crocus of Mars commonly known as rust) that could be easily scraped off.

The Crocus of Mars was then made into a tincture that could be used to treat certain medical conditions. To render the rust amenable to consumption, it had to first be prepared by a process of absorption ("imbibition") and extraction ("decoction"). In "The Crocus of the Metals, or the Tincture," Paracelsus stated:

> *The consumption of the rust is made by the imbibition of those things which produce rust, and by a decoction extracting the color of rust. Take old Urine poured away from its deposit, several cups of it, in which dissolve three handfuls of ground salt. When you have strained it, boil it and skim it carefully. In this again dissolve a handful of bruised vitriol, with two or three ounces of bruised sal ammoniac, and then carefully skim again. With this liquid imbibe some filings, and boil until it can be pulverized. The dust thus produced reverberates over a powerful fire, continually stirring it with an iron rod, until it changes from its own color to another, and at last into the hues of most brilliant violet. From this you can easily, with spirits of wine [alcohol] or distilled acetum, draw off the tincture, and when it is extracted by separation of the elements you will collect what remains at the bottom of the glass, by means whereof you will be able to produce wondrous effects, both within and without the body.*[17]

The iron oxide must be absorbed into a paste or gel (imbibition). First a mixture was made of urine, salt, vitriol (presumably iron sulfate or possibly copper sulfate), and sal ammoniac (in regard to vitriol, "bruised" could either be a reference to the coloration which would be a greenish color for iron sulfate or a bluish color for copper sulfate, or could be a reference to the mechanical treatment of pulverization. Given the additional association of "bruised" with sal ammoniac, the most appropriate understanding is that a mechanical action was involved). Iron filings were then added to this mixture, which was boiled until a residue resulted. After mixing the pulverized residue with alcohol or vinegar, followed by distillation, the tincture could be drawn off and used to treat certain medical conditions.

CHAPTER 6

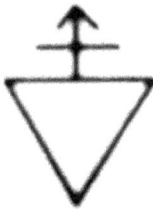

Crocus of Mars

One of the uses for the Tincture of Mars, according to Paracelsus, was to stop the flow of blood from wounds. He pointed out, for example, that Crocus of Iron (burnt iron) was often used by iron workers to cauterize their injuries. If it was reduced by the reverberatory into alcohol, Crocus of Iron could also heal ulcers (which could also be cured by Oil of Iron or Oil of Mars, as it was also known)[18] The preparation of Oil of Mars involved simply grinding Crocus of Mars into a moist, fine powder, washing the powder, and then separating the powder that had precipitated out. After drying, the powder could be mixed with egg yolks into a paste that was allowed to dry and once again beaten into a powder. Finally, this concoction was spread on a glass slab and placed in a cool place such as a wine cellar. The clear oil that resulted, the Oil of Mars, Paracelsus used to treat ulcers.[19] He called Oil of Iron *"ferrum potabile"* and advocated its use in his work *De Tumoribus et Pustulis Morbi Gallici,* which purported that it could cure "... every genus of humid, fluid, and flaccid ulcers...."[20]

Basil Valentine (15th or 16th century CE)
Paracelsus's thoughts pertaining to iron appeared to have carried over into Basil Valentine's views of the metal. The language in Valentine's *Twelve Keys* was strikingly similar to Paracelsus's, indicating that either Valentine had borrowed from Paracelsus or possibly that they both acquired their theories and descriptions from a prior common source. (The identification of authorship and assignment of date to the works attributed to Valentine has complicated this issue. Basil Valentine was possibly a 15th-century alchemist and Canon of the Benedictine Priory of Saint Peter in Erfurt, Germany. More likely, however, his name was a pseudonym for one or more 16th-century German authors. If this is accurate, then, the unknown author or authors would have been well aware of Paracelsus's work.)

In the "Ninth Key" of his *Twelve Keys,* Valentine personified Mars, the Roman god of war, as the head of military affairs in the royal court of the King of the Metals, Gold, who executed his office with a fiery zeal or passion.[21] Interestingly, Valentine further embellished Mars's place at court with a standard bearer known as Geometry who carried before him a crimson banner. The color was fitting since crimson, in addition to being a purplish red color, could also be understood to be sanguinary (characterized by bloodshed). It is unclear why Valentine also chose to personify Geometry, however. Perhaps, since the word geometry comes from the Greek meaning a measuring of earth (Greek "*geo*"? for earth and "*metria*" for the measuring of), the knowledge of geometry would have been useful to farmers and landowners. Could it have been that since Mars was also an agricultural deity, Valentine associated geometry with Mars? Another association could have come from the use of the instruments of medieval warfare such as battle axes, swords, and the catapult. Use of the catapult would have required calculation of trajectories and distances. (Regardless of how Valentine derived his description, his initial treatment of Mars was certainly a colorful one.)

According to Valentine, Mars, like all the metals, originated from a "... supernatural, flying, fiery Spirit which preserved itself in the Air, seeking its habitation naturally in the Earth and Water, wherein it could rest and operate...." He attributed the nature or character of the metals to how this Spirit interacted with the bodies of each. For instance, the inertness of gold, he asserted, was because it was so tightly packed it lacked the pores which allowed the fiery Spirit to penetrate. The other metals were not so compacted, permitting the Spirit to possess them. Valentine called this Spirit the Tincture of Gold. But because of the reactivity (inconstancy) of the other metals, this Tincture had to depart from them. Presumably, the alchemist had to find the means by which the Tincture of Gold, which was trapped within the metals, could be extracted to produce a "tinging Spirit." The best metals for this extracting process, according to Valentine, were Mars (iron) and Venus (copper):

> *Know that the first Tincture and Root of all Metals, is likewise a supernatural, flying, fiery Spirit; which preserves itself in the Air, seeking its habitation naturally in the Earth and Water, wherein it can rest and operate: This Spirit is found in all Metals, more abundant in other Metals than in Gold, because Gold, by reason of its well digested, ripened, and fixt body, is tight, close, and compact, and therefore no more can enter into its body than is just requisite; but the other Metals have not such fixt bodies, for their pores are open, and far extenuated, therefore the Tincture Spirit can the more abundantly pass*

> *through and possess them. But because the bodies of the other Metals are inconstant, the Tincture cannot remain with those inconstant bodies, but must depart. And whereas the Tincture of Gold is found in none more plentiful than in Mars and Venus, as Man and Wife, their bodies therefore are destroyed, and the tinging Spirit taken out of them, which makes Gold sanguine....*[22]

Using the same language and line of thought as Paracelsus, but reordering the constituents in contrast to Paracelsus, Valentine considered Mars to be composed primarily of salt. Sulfur and Mercury then followed, in that order. This constitution, according to Valentine, resulted in Mars being characterized as the hardest, strongest, least gentle, and grossest of the metals. (Since sulfur was the principle for congealing or "hardening" the mercury, he concluded sulfur led to great hardness.):[23]

> *... the metal of Mars being ordained in its degree by a gross salt before others in the greatest quantity, is found to have the hardest, ungentle, strongest, and grossest body, which nature appropriated and granted to it, it hath the least portion of mercury, but more of sulphur, and most of salt....*[24]

In spite of the not-so-flattering description of Mars, Valentine affirmed that iron, properly prepared, had certain stimulating effects. When mixed with wine, the spirit of Mars acted as an energizing tonic, enhancing one's heart, courage, and senses:

> *... if you can know the right and true spirit of Mars... that one grain of its spirit or quintessence drunk with the spirit of wine, strengthens the heart, courage, and senses, so that you shall fear no foes; it raises up in him the courage of a lion, provokes a desire to hunt and fight at Venus sports. When the conjunction of Mars and Venus are rightly placed in a certain constellation, they bring fortune and victory in love and affection, in battle and joy....*[25]

So Valentine remained consistent with his portrayal of iron—in the guise of Mars—by attributing the characteristics of the god of war to the effects of his concoction of the Spirit of Mars and wine.

Martinus Rulandus (1532–1602)

As a lexicographer of Paracelsian terms, Rulandus provided additional names and descriptions for iron, which he listed under the Latin "*Ferrum*." He defined "*Ferrum*" as a metal of livid color, but also containing a certain quality of redness along with an impure whiteness which possibly described one of the ores of iron. Curiously, he discussed the composition of iron using only the principles of Mercury and Sulfur, unlike Paracelsus who had also included Salt. Thus, Rulandus attributed the constitution of iron only to the presence of fixed, earthy Sulfur and fixed, earthy quicksilver with the Sulfur predominating. The result, according to Rulandus, was accorded the name Mars by "the chymists."

Under the entry "*Ferrum*," Rulandus also provided related terms. Sideritis referred to iron filings or Scale of Iron, and sometimes was used as the term for the magnetic form of iron or magnetite. From Dioscorides, Rulandus listed Recrement of Iron, also called Scoria (iron slag) and Excrement of Iron, also called the stone Sideritis or the German Thunderstone (Aerolite). Miscellaneous terms included: Ferri Scons—a reference to iron filings, Ferrum Indicum—a reference to steel, and Ferrugo—a reference to rust, scoria, iron refuse, must, or iron mold.[26]

Iron Filings

Eirenaeus Philalethes (an alias of George Starkey, 1628–1665)

Perhaps one of the most obscure alchemical writings about iron has come from George Starkey, writing under the pseudonym of Eirenaeus Philalethes. In his *An Exposition Upon Sir George Ripley's Preface*, iron was part of a "Trinity of Substances," also known as "our true Fountain," consisting of "three Springs"— "a Water or Mercurial Bond," the "Blood of our Green Lyon" (meaning that it was green in the sense that it lacked Sulfur and was completely volatile, which was most likely crude Antimony), and a "Spirit" or "Chaos," which was useful in human affairs, a reference to common iron:[27]

> *This Fountain hath three Springs, and these are three Witnesses which testifie to the Artist of the truth of his proceedings; these are*

> the Spirit, the Water, and the Blood, and these three agree in one; the Water is a Mercurial Bond, which the Sophisters can behold so far as the outward shell reacheth, but the wise man can behold his hidden secret Centre: the Blood is of our Green Lyon, which is indeed the greenest or rawest of the three: for it hath no manner of Metalline Sulphur, not a grain, and therefore is Totally Volatile, and it is more raw than the common Water, and yet it is called the Blood, for a most secret reason, because it is the seat of the Life, which is the Spirit, as Blood in man is the seat of his Life; yea the Spirit by this Soul of our Green Lyon, is made manifest, and is united to it, so that though it be very green or unripe, yet that inhabits it, which is both pure and ripe, and can and will digest it with the Water, and make both become life with life: Now the Spirit is nothing else but a Chaos, the Wonder of the Wonders of God, which every man almost hath, and knows it not, because as it appears to the World it is compact in a vile despised form; yet is it so useful, that in humane Affairs none can want it....[28]

In Philalethes's *Introitus Apertus ad Occlusum Regis Palatium*, "Chaos" referred to antimony. In chapter II of the *Introitus*, however, Philalethes considered the "Chaos" to be both Magnes and Chalybs, that is, magnet and steel:

> The components of our water [Mercury of the Sages] are fire, the vegetable "Saturnian liquid," and the bond of Mercury. The fire is that of mineral sulphur, which yet can be called neither mineral nor metallic, but partakes of both characters: it is a chaos or spirit, because our fiery Dragon, that overcomes all things, is yet penetrated by the odor of the Saturnian liquid, its blood growing together with the Saturnian sap into one body which is yet neither a body (since it is all volatile) nor a spirit (since in fire it resembles melted metal). It may thus be very properly described as Chaos, or the mother of all metals. From this Chaos I can extract everything—even the Sun and Moon—without the transmutatory Elixir. It is called our Arsenic, our Air, our Moon, our Magnes, and our Chalybs....[29]

According to William Newman (Prof. of the History of Science, Indiana University), in his work *Gehennical Fire*, antimony was the Chaos, the mother of all metals, since it was a body that displayed metallic and volatile qualities.

Further, the reactants that produced this "Chaos" were Saturnia (crude Antimony or Antimony sulfide) and the "fiery dragon," which was iron and also called "mineral sulfur." From Paracelsian theory, the principle of Fiery Sulfur was believed to reside in iron in great quantity, evidenced by iron's relatively high melting point. Crude antimony, even though Philalethes described it as containing impure sulfur, did not contain metallic sulfur, which had to be obtained from iron. In other words, iron reduced antimony sulfide to yield metallic antimony and iron sulfide. The sulfur obtained from iron was also seen as a "fiery dragon" because of its volatility (again, alchemical imagery seems to have been conflated).

Using the language of Philalethes, the reduction reaction, then, would be described in this way: The "Magnes" was crude antimony that attracted the spirit (principle) of Sulfur from the iron and joined it with the Mercury (Paracelsian principle) of the antimony to form metallic antimony, and the "Chalybs" was the ferrous spirit of the iron attracted by the "Magnes" (crude antimony).[30] In other words, the crude antimony reacted with the iron to form metallic antimony.

Nicholas Lemery (17 November 1645 – 19 June 1715)

In stark contrast to George Starkey's alter-ego, Lemery was more transparent in defining his terms and explaining the uses for iron. In agreement with the alchemists and chymists who came before him, Lemery described iron as a porous metal consisting of "Vitrolick Salt, Sulphur, and Earth." Because of its porous nature, iron was subject to the "Impressions of the Air, and, consequently, to contract Rust." Lemery properly noted that iron readily reacted with air to form rust or what he called Saffron of Mars, for which he also provided several recipes, some of which used acids. To understand the effect of acid on iron, Lemery provided a theory as to why iron reacted with acids. Although Mars contained Vitriolic Salt, he explained, it retained its alkali nature because it reacted with acids. His explanation was that iron was more Earth than Salt, which provided sufficient pores, both in size and quantity, to receive the points of the acid. Here was a clear application of Lemery's theory of acids and alkalis that conceived acids as having points, analogous to cogs, and alkalis as having pores receptive to the points of the acid. Since the pores of the alkali were properly disposed, the acid was able to penetrate the body of iron, separating ". . . whatever stands in (the acid's) way . . . ," and, consequently, dissolving the iron.

Saffron of Mars, then, became a starting point for three substances derived from iron: Salt or Vitriol of Mars, Tincture of Mars, and Mars Diaphoretick or Flowers of Steel (also known as the Martial Flowers of Sal Armoniack). Salt or Vitriol of Mars was simply iron reduced to the form of salt (Lemery's term) using alcohol and sulfuric acid:[31]

> This preparation is an iron opened and reduced into the form of salt by an acid liquor. Take a clean fryingpan and pour into it an equal weight of spirit of wine and oil of vitriol. . . .[32]

Tincture of Mars was produced by the dissolution of iron using acid of tartar (tartaric acid), and Mars Diaphoretick was obtained by the sublimation of iron particles with volatile salts such as sal armoniack (ammonium chloride).

These three compounds were used to treat various ailments, according to Lemery. Mars Diaphoretick when mixed with spirit of wine (alcohol) was sudorific, causing sweating, and aperitive, stimulating the appetite. It was believed to be effective for treating malignant fevers, lethargy, palsy, scurvy, asthma, purification of the blood, and stoppage of looseness of the bowels and vomiting. Tincture of Mars was also seen as an aperitif and as a treatment for severe weight or muscle loss (cachexia), and dropsy. Finally, Vitriol of Mars was thought to be good for any disease that "proceeded from obstructions."[33]

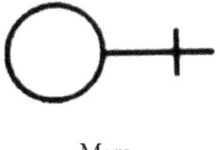

Mars

Conclusion

Though iron was seen as the least amenable metal to transmutation into a more noble metal, it was still held in high regard because of its strength and utility. Appropriately, Mars came to be associated with the metal of choice for weapons of war and for implements of agriculture, since the god of the red moving star ruled over farmlands, harvests, and their defense. In addition, medicaments, derived from iron, were used to treat such conditions as wounds and lethargy, providing a healing balm and invigorating the person—treatments most necessary for working and defending one's home and land.

Further, the experiments and discoveries of the alchemists in regard to iron allowed them to refine their theories of matter and explain the observed differences among the seven metals.

Notes

1. J.E. Cirlot, *A Dictionary of Symbols*, 2nd Edition, (New York: The Philosophical Library, 1971), 204.
2. Symbols for Mars/Iron are from Philip Wheeler, *Alchemical Symbols, Fourth Edition, R.A.M.S. Library of Alchemy, Vol. 21*, (Kansas City, MO: R.A.M.S. Publishing Company, 2018), 33, 47, 54.
3. Eric John Holmyard, *Makers of Chemistry*, (Oxford: Clarendon Press, 1931), 4.
4. John Emsley, *Nature's Building Blocks: An A-Z Guide to the Elements*, (Oxford: Oxford University Press, 2011), 258.
5. Raphael Patai, *The Jewish Alchemists: A History and Source Book*, (Princeton, NJ: Princeton University Press, 1994), 104-105.
6. Ibid.
7. Dark, bluish, grey. "Whiteness" probably refers to a silvery color or appearance.
8. E. J. Holmyard and Richard Russell, *The Works of Geber*, (Kessinger's Legacy Reprints, Kessinger Publishing), 138–139.
9. Ibid., 68–69.
10. Ibid., 138–139.
11. Ibid., 146–152, 156–157.
12. Paracelsus, *The Hermetic and Alchemical Writings of Aureolis Philippus Theophrastus Bombast, of Hohenheim, Called Paracelsus the Great*. Edited by Arthur Edward Waite in two volumes, (Mansfield Centre, CT: Martino Publishing, 2009), Vol. I, 7, 7n. See also *De Elements Aquae*, Lib. IV., Tract III.
13. Ibid.
14. Ibid., 5, note.
15. Ibid., 14.
16. Ibid.
17. Ibid., Vol. I, 199.
18. Ibid., Vol. I, 200n.
19. Ibid., Vol. II, 140.
20. Ibid., Vol. II, The Archidoxies of Theophrastus Paracelsus, "Concerning the Specific Matrix," 67.
21. Basilius Valentinus, *Twelve Keys of Basilius Valentinus, Also Includes Of Natural and Supernatural Things, The R.A.M.S. Library of Alchemy, Vol. 1*, (Stuarts Draft, VA: R.A.M.S. Publishing Co., 2015), 81.

22. Ibid., Chapter II, *Of Natural and Supernatural Things*, 135-136.
23. William R. Newman, *Gehennical Fire: The Lives of George Starkey, an American Alchemist in the Scientific Revolution* (Chicago: The University of Chicago Press, 1994), 129.
24. Valentinus, 163-165.
25. Ibid.
26. Martinus Rulandus, *A Lexicon of Alchemy, Containing a Full and Plain Explanation of All Obscure Words, Hermetic Subjects, and Arcane Phrases of Paracelsus* (Zachariah Palthenus, bookseller, in the free republic of Frankfurt, 1612), 140-144.
27. William R. Newman, 130-131. See also Eirenaeus Philalethes, *An Exposition Upon Sir George Ripley's Preface* (London, Printed for William Cooper at the Pellican in Little Britain, MDCLXXVII), 28-29. (umich.edu). Retrieved 14-Oct-2023.
28. Eirenaeus Philalethes, *An Exposition Upon Sir George Ripley's Preface*, 28-29.
29. Eirenaeus Philalethes, "Chapter II, Of the Component Principles of the Mercury of the Sages," *Introitus Apertus ad Occlusum Regis Palatium*. Levity.com. Retrieved 14-Oct-2023.
30. William R. Newman, 126-130.
31. Nicolas Lemery, *A Course of Chymistry, The Fourth Edition*, London, 1720, (Published by Creative Media Partners on behalf of Gale Research, Inc.), 108-130.
32. Lemery, 121.
33. Lemery, 122-130.

Chapter 7

Copper, An Earthy Venus

Introduction

Copper and its associated goddess-planet, the Roman Venus, are ancient. Possibly, they shared the same location of origin. Venus, the syncretized form of Aphrodite, was the goddess of beauty, fertility, procreation, passion, and love. A very ancient deity, her origins have been traced back to the island of Cyprus. According to Hesiod's *Theogony,* Aphrodite was born off the coast of Cyprus. She had many epithets, including Aphrodite Urania (Heavenly Aphrodite) and Aphrodite Pandemos (Aphrodite for all the people). She was also known by titles that situated her origin—Cypris (Lady of Cyprus) and Cytherea (Lady of Cythera). Both locations claimed to be her place of birth. It was on Cyprus that the Greeks first encountered her cult.[1]

Venus[2]

From ancient times, Cyprus was known for its copper. Though the metal has been known for at least 10,000 years (Copper beads were discovered in northern Iraq to be older than 10,000 years), the metal was first smelted about

5000 years ago in Cyprus,[3] which was either the origin of the name of the metal or which was named after the metal. One of the earliest names for the metal was aes cyprium (metal of Cyprus), later corrupted to the Latin cuprum, from which the English word was derived.[4]

Venus and copper, however, shared more than a common origin. After gold and silver, copper was the most lustrous and desirable of the metals; it was used to make tools, fine ornaments, and jewelry. Venus shared these qualities, being the brightest of the "moving stars" after the Sun and the Moon, and the most desirable of the ancient goddesses.[5]

Early smelting practices involved extracting the metal from malachite, an easily identifiable greenish ore of copper—copper carbonate ($Cu_2CO_3(OH)_2$). Sometime after 5000 BC (probably in the period between 4000-2000 BC), it was discovered that copper could be greatly strengthened and honed to a fine cutting edge if alloyed with tin. The alloy, known as bronze, became the metal of choice for weapons and tools until the beginning of the Common Era when another alloy of copper was discovered—brass, an alloy of copper and zinc.[6]

Because copper had an aesthetic and a practical beauty, knowledge of it was desired by metallurgists, naturalists, and alchemists alike. Thus, over the course of time, many names, descriptions, and applications of Venus's metal arose to express insights gained about the metal that stood between the Sun and the Moon, and was considered to have a special relationship to Mars.

Pliny the Elder (23-79)
Pliny affirmed that copper was first discovered on Cyprus and that the metal was extracted from a "stone" called chalcitis[7], which was composed of copper, misy (copper pyrites), and sory (iron sulfide, possibly CuS or CuS_2). A better copper ore, according to Pliny, was known as "aurichalcum," the exact composition of which is unknown. It may have been a native brass—a mixture of copper and zinc. The edition of *Natural History*, which was edited by John Bostock, M.D., F.R.S., and H.T. Riley, Esq., B.A., indicated that "aurichalcum" may have been originally written "orichaleum," meaning "mountain brass." "Orichaleum" was also mentioned in Plato's *Critias* and was considered to be second only to gold, with which Pliny agreed:

> *In Cyprus, where copper was first discovered, it is also procured from another stone, which is called "chalcitis." This, however, was afterwards considered of little value, a better kind having been found in*

other regions, especially that called "aurichalcum," which was long in high request, on account of its excellent quality....[8]

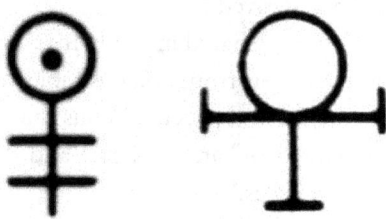

Symbols of Orichaleum or Brass[9]

Pliny identified three different kinds of copper: "coronarium," "regulare," and "caldarium." Both "coronarium" and "regulare" were ductile and malleable. If the former were mixed with gold, it took on a fiery red color and was known as pyropus (i.e., sparkling like fire). The third kind was called "caldarium" (a cast bronze) and, unlike "coronarium" and "regulare," would break when struck with a hammer:

> *We will now return to the different kinds of copper and its several combinations. In Cyprian copper, we have the kind known as "coronarium," and that called "regulare," both of them ductile. The former is made into thin leaves and, after being colored with ox-gall, is used for what has all the appearance of gilding on the coronets worn upon the stage. The same substance, if mixed with gold in the proportion of six scruples of gold to the ounce and reduced into thin plates, acquires a fiery red color and is termed "pyropus." In other mines again, they prepare the kind known as "regulare," as also that which is called "caldarium." These differ from each other in this respect that, in the latter, the metal is only fused and breaks when struck with the hammer, whereas the "regulare" is malleable or ductile....*[10]

In addition to cataloging various items made from brass and bronze, Pliny listed the ailments and diseases for which copper compounds were effective. He claimed that calcined copper acted as a desiccative, healed wounds, arrested discharges, acted as a detergent on incrustations of the eyes, and could be used for white spots and cicatrizations (healings by scar formation) on the eyes.[11]

Flower of Copper (possibly red cuprous oxide) and Scoria of Copper were also used for medicinal treatments. The difference between the "flower" and the "scoria" was that the former consisted of cuprous oxide—copper (I) oxide, while the latter consisted of the metal, mixed with heterogeneous matter, that had been separated during the melting of the ore (i.e., the slag). A final copper compound was known as "stomoma." Both the "stomoma" and the Flower of Copper were procured by the calcination of copper in shallow earthen or brazen pans. In the case of the Flower of Copper, the calcined copper was removed to another furnace where repeated application of hot air made the metal separate into small scales. Lepis, an associated product, was produced by hammering the calcined cakes of copper. Flower of Copper, on the other hand, fell off spontaneously. "Stomoma," Flowers of Copper, and lepis could all be used for excrescences in the nostrils and in the anus, and for dullness of the hearing (being forcibly blown into the ears through a tube). Finally, they could also be applied to swellings of the uvula, and, if mixed with honey, to swellings of the tonsils.[12]

Unfortunately, Pliny confused the copper alloys. He did not seem to recognize that aurichalcum was the same product as that made by mixing cadmia and molten copper—"That which is at present held in the highest estimation" he wrote, "is the Marian, likewise known as the Corduban: next to the Livian, this kind most readily absorbs cadmia, and becomes almost as excellent as aurichalcum. . . ." Besides, his indiscriminate use of the term Aes for copper, bronze, and brass also clouded the issue.[13]

Al-Khwārizmī (Muḥammad ibn Mūsā al-Khwārizmī, d. ca 997)

One of the best of the early taxonomies of substances and apparatus in the alchemical writings came from Al-Khwarizmi. In his encyclopedic work entitled, *The Keys to the Sciences* (written between 976 and 980 CE), he organized his taxonomy into three divisions: apparatus, substances (naturallyoccurring and artificial), and operations. Under section two of "The Ninth Chapter of the Second Discourse: On Alchemy," Al-Khwarizmi categorized copper as one of the bodies, which also included gold, silver, iron, lead, tin, and quicksilver (mercury). Within this listing, he also attached the planetary association for each metal—Sun for gold, Moon for silver, Venus for copper, Saturn for lead, Mars for iron, Jupiter for tin, and Mercury for quicksilver. After the bodies, he listed spirits, salts, vitriols, and a variety of minerals. He also included substances produced artificially.

In particular, he described one of the artificially produced substances related to copper—verdigris. A copper salt of acetic acid, verdigris, according to Al-Khwarizmi, was produced from immersing copper sheets in vinegar (acetum). When the sheets turned green, the verdigris was then scraped off.[14] His instructions were slightly vague concerning the color. By green, did Al-Khwarizmi mean dark green, light green, bluish-green, or some other hue in that range of colors? Today, verdigris is a collective term for copper acetate, whose chemical varieties produce different hues. Since verdigris actually ranges in color from dark green to bluish-green, depending on the hydration level and degrees of basicity, the actual chemical composition of Al-Khwarizmi's verdigris could be one of several variations of copper acetate—from copper(II) acetate monohydrate $(Cu(CH_3CO_2)_2 \cdot (H_2O))$, which is a dark green, to $Cu(CH_3CO_2)_2 \cdot CuO \cdot (H_2O)_6$, which is known as "blue verdigris."[15]

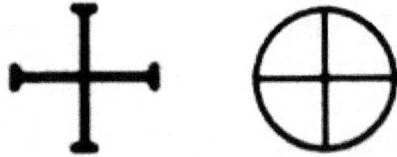

Acetum (Vinegar), and Verdigris (Green of Copper, Virido Acis)[16]

Albertus Magnus (Albert the Great, 1193–1280)
Though Albertus Magnus has been seen as one of the "greats" among the early alchemists, his information on alchemy was largely taken from Avicenna's work, *De Anima*. But much of Avicenna's knowledge came from such sources as al-Razi. Thus, Albert Magnus's writings have presented a consistent, longline of thought on many alchemical matters. Preserving the Aristotelian tradition of the four elements—Fire, Air, Water, and Earth—and the four qualities—Hot, Cold, Wet, and Dry,—Magnus explained that the metals were "formed from water congealed by intense cold and dryness."

He also applied the Jabirian "Mercury-Sulfur" theory (the theory that metals consisted of two principles—Mercury and Sulfur), affirming that Mercury (Argentum vivum or hydrageros) was the principle of metals... of which there are seven: lead, tin, iron, bronze (Aes), brass (Aurichalcum), copper (Cuprum), silver, and gold. ("... Electrum not being a simple metal but made by art.") Strikingly, the actual number of metals listed was nine, not seven. Since bronze and brass were alloys of copper, Magnus most likely considered these as variations of copper. For example, Book IV of his *De Mineralibus* mentioned the preparation of brass by melting copper with calamine (derived from *lapis calaminaris*, a Latin corruption of Greek cadmia (καδμία), the name for zinc ores) or tutia (an oxide of zinc) with glass as a flux, which "reflects the vapor" into the copper.[17] Clearly, he understood brass as an alloy of copper.

Pseudo-Geber (Paul of Taranto, 13th century)
Paul of Taranto's *Of the Sum of Perfection* described copper as a livid, green metallic body:

> *Chapter XII, Of Venus or Copper, Part III, Book I, Of the Sum of Perfection:*
> "Our intended discourse now is of Venus, or Copper. It is a metallic body, livid, partaking of a dusky redness ignoble (or sustaining ignition), fusible, extensible under the hammer, but refusing the cupel and cement. Therefore, Venus (as is declared) in the profundity of its substance pretends to the color and essence of gold, and it is hammered being heat red-hot, as silver and gold is. Therefore, hence you may learn a secret: for it is the medium of Sol and Luna, and easily comes to convert its nature to either; and it is of good conversion and of little labor. It agrees very well with Tutia, which citrinizes (or colors) it with good yellowness; and hence you may reap profit. For we are excused by it from the labor of induration (or hardening) and ignition of it. Therefore, take it before all other imperfect bodies, in the lesser and middle work, but not in the greater. Yet this hath a vice beyond Jupiter, viz. that it easily waxeth livid and receives infection from any sharp and acute things; and to eradicate that is not an easy but a profound art."[18]

Pseudo-Geber may have described an ore of copper since pure copper has an orange-red color. The most common copper ore, chalcopyrite, has a golden yellow color, but may also have a purplish tarnish that could be described as livid. (This discoloration, most likely, results from the oxidation of the chalcopyrite, which, when exposed to air, tarnishes to a variety of oxides, hydroxides, and sulfates.) In addition, copper shared the color of gold, and on heating was malleable like gold and silver. Unlike the noble metals, however, copper does not purify as gold and silver do. Refusing the cupel, in comparison to gold and silver, possibly indicated, to Pseudo-Geber, that copper did not separate from its ore into a pure form as gold or silver did under cupellation—a process based on the principle that precious metals do not oxidize or react chemically, unlike base metals.[19] When the ores of precious metals are heated at high temperatures, the metals separate out, and the remaining material forms slags or other compounds. Also, according to Pseudo-Geber, copper would not respond well to cementation (an alchemical process in which layers of the metal would be mixed with a powdered, corrosive material and then welded together by heating to high temperature).[20] Other texts, however, seemed to contradict his statement here regarding the cementation of copper (see the excerpt below from *Of the Invention of Verity or Perfection*, Chapter XV, *Of the Preparation of Venus*, and the following discussion). If by cementation, he meant the use of cements (a paste of substances) by which copper combined with the substances of the paste and thereby separated from the ore, then his usages of the cementation process would be consistent. Most likely, he may have only meant that copper was less amenable to cementation than gold or silver. He stated later, in the discourse "Of Cement" in the *Summa*:

> *We have said that some bodies more, and others less, are burned by the calcination of fire, viz., they which contain a greater quantity of burning sulphur, more; but they that contain less, less. Therefore, seeing Sol hath a less quantity of sulphur than other metallic bodies, it is not (in the midst of all mineral bodies) burnt by inflammation of fire. And Luna, next to Sol, partakes of a less quantity of sulphur than the other four bodies, yet more than Sol. Therefore, according to this, it can less bear the ignition of inflammation for a long space of time than Sol; and by consequence less bear things burning by a like nature, but Venus less than it; because it consists of more sulphur, and of greater earthiness, than Luna; therefore can less bear the inflammation of fire.*[21]

Copper, however, alloys well with tutia (ZnO, or $ZnCO_3$), which would be mixed with molten copper and charcoal. The gases would be given off and would leave brass. Since brass was harder than pure copper, the alchemist could eliminate the effect of hammering the copper in fire and instead harden it by simply combining it with tutia.[22]

Further, Pseudo-Geber pointed out that copper was very porous at a microscopic level because it contained heterogeneous earthy particles that interrupted the packing of its mercury particles:

> *Of Venus and Mars, the way of Calcination is one; yet diverse from the former, by reason of the difficulty of their liquefaction. And it is this, either of these bodies reduced into plates, must be heated red-hot, but not melted. For by reason of the great quantity of earthiness in them, and the large measure they have of adustive and flying sulphureity, they are easily this way deduced into calx. And that therefore is, because of the reason of much earthiness, mixed with the substance of argentivive, the due continuation of argentivive is disturbed. Therefore porosity is caused in them, through which the sulphureity passing may fly away; and the fire, by that means having access to it, burn and elevate the same. Whence it comes to pass, that the parts are made more rare, and through discontinuity of the rarity converted into ashes. The experience of this is manifest, because plates of copper exposed to ignition yield a sulphureous flame, and cause pulverizable scales in their surfaces. And that therefore is, because from the parts more nigh, a more easy combustion of sulphur must necessarily be made.*[24]

The large amount of sulfur-bearing earthy particles disrupted the continuity of the Argentvive (quicksilver) particles, creating porosity—". . . because of the reason of much Earthiness, mixed with the Substance of Argentvive, the due continuation of the Argentvive is disturbed." The spaciousness of the porosity, then, allowed the sulfurous quality to escape, ignite, and burn off. This action produced finer particles that converted into ash or "Calx" (". . . the Parts are made rarer, through Discontinuity of the Rarity converted in Ashes (calcined) . . .").[25]

Finally, in his general description, Pseudo-Geber pointed out copper's propensity to corrode—"Yet this hath a Vice beyond Jupiter, viz. that it easily waxeth Livid, and receives Infection, from any sharp and acute things. . . ." That is, copper is worse than tin, discoloring ("waxeth Livid"—a reference to an oxide of copper) and reacting with air ("sharp and acute things"—a reference to air and vapors) to form superficial oxides.[26]

Pseudo-Geber also discussed the nature of copper using the concepts of the Mercury-Sulfur Theory. If fixed sulfur was present to a greater extent than mercury (Argentvive), then copper would result. His proof was that when copper was heated, a "sulphureous" flame would arise, which would be the unfixed (volatile) sulfur given off. After repeated heating, the amount of the unfixed sulfur would be lost, leaving only the fixed sulfur that would inhibit copper's fusion, and would also explain copper's hardness. Both the inhibition to fusion and the hardness were signs of the greater amount of fixed sulfur, according to Pseudo-Geber.

> *Chapter VIII, Of the Nature of Venus or Copper, Part I, Book II:*
> *"Wherefore, returning to our purpose, we say that if the sulphur be unclean, gross, and fixed, as to its greater part; but as to its lesser part not fixed, red, and livid; in relation to the whole, not overcoming, nor overcome; and this fall upon gross argentive, copper must necessarily be created thereof. The probation of all these is easier, by things given from the nature of them. For when copper is exposed to ignition, you may discern a sulphureous flame to arise from it, which is a sign of sulphur not fixed. And the loss of the quantity of it by exhalation, through the frequent combustion thereof, signifies that it hath fixed sulphur. For from that is caused the slowness of its fusion, and induration (or hardness) of its substance, which are signs of the multitude of its fixed sulphur. And that there is sulphur red and unclean, conjoined with unclean argentive, is known by sense . . ."* [27]

In addition to a description of the qualities and nature of copper, Pseudo-Geber also offered, in both the *Summa Perfectionis* and *Of the Invention of Verity or Perfection*, various preparations of the metal that could be later used for producing medicinal treatments.

> *Chapter V, "Of the Preparation of Venus," Part II, Book II, Summa Perfectionis:*
> *"Therefore, imitating the Order of the premised, We declare the Preparations of two Bodies likewise. But first of Venus, afterward of Mars. The way of the Preparation of Venus is manifold. One way is by Elevation, but another is completed without Elevation. The way by Elevation is that Tutia be taken, with which Venus well agrees, and that it be ingeniously united with it. Then it must be put in its Vessel of Sublimation to be sublimed, and by a most excelling degree of Fire its more subtile part elevated; which will be found of most*

bright Splendor. Or it may be mixed with Sulphur, and then elevated by its now mentioned way of Elevation. But without Sublimation, it is prepared, either by cleansing Things, in its Calx, or in its Body; as by Tutia, Salt, and Allomes, or by a Lavament of Argentvive, the way of which We have given: or else by Calcination and Reduction of that, which is dissolved into the Nature of Body; or (as We said) it is cleansed by a Lavament of Argentvive, as all other Bodies diminished from Perfection are." [28]

Chapter XV, "Of the Preparation of Venus," Of the Invention of Verity or Perfection:
The Purgation of Venus is twofold, one for the White, and the other for the Red. For the White, it is thus: Calcine Venus with Fire only, as We shewed in our Sum of Perfection. R. Venus thus calcined, grind, 1 lib. Of it with four Ounces of Arsenic sublimed, and imbibe the Mixture three or four times with Water of Lithargiry, and reduce the whole with Salt-Peter, and Oyl of Tartar; and you will find the Body of Venus white and splendid, fit for receiving its Medicine.

    For the Red, the Preparation of it is thus: Grind one pound of Filings of Venus, with four ounces of Sulphur; or Cement Plates of Copper with Sulphur, and so calcine; and wash the Calcinate with the Water of Salt and Allom, and then (with Things reducing) reduce into a clean Body apt for the Red Tincture.[29]

The last preparation called for the cementation of copper and sulfur, which may seem contradictory to his description of copper. However, copper readily reacts with sulfur to form copper sulfide, and so the exposure of a copper plate to sulfur under heat would produce calcined copper sulfide which would be a red powder.

Paracelsus (c. 1493 – 24 September 1541)
In *De Elemento Aquae*, Paracelsus asserted that Venus was the first metal generated by the Archeus of Nature, the dispenser of things. According to Martinus Rulandus, the lexicographer of Paracelsian terms, Archeus was the divider of the elements, disposing of them and relegating places, genera, and species to all things (In other words, the Universal Agent specialized in each individual thing... that sets all nature in motion).[30] Considered the most exalted invisible

spirit/virtue hidden in nature, the Archeus generated the metals from the three prime principals (Mercury, Sulfur, and Salt) after the marcasites (white iron pyrite—iron sulfide) and cachimaiae (imperfect bodies)[31] have been separated from these principles: "It (Venus) is formed from the gross redness which is purged off from the primal Sulfur, of the light red expelled in like manner from the Mercury, and of the deep yellow separated in the purification of the prime Salt by this same Archeus."[32] (Paracelsus can be confusing. Here, in *De Elemento Aqua,* he seems to be saying that the metals can be separated by a "Universal Agent" of nature. Yet, in *A Book about Minerals,* he says that the separation of the metals from their impurities can only be accomplished through alchemy[33]— which implies the necessity of human effort. What he wrote in *Labyrinthus medicorum errantium* (written 1537-1538) may clarify this inconsistency. In that work, he advocated that the term "alchemy" applied to all transformation processes in nature as well as those brought about by human hands. With this understanding of alchemy, then, metals could be separated from their impurities both by the Archeus and the alchemical adept.) [34]

Paracelsus considered the transformations of all the metals into their oxides, sulfides, sulfates, etc., as mortifications. He saw that the death of all natural things was nothing but an alteration into a new and different nature. Thus, "... the death or mortification of the metals is the removal of their bodily structure, and of the sulphurous fatness which can be removed from them in many ways, as by calcination, reverberation, resolution, cementation, and sublimation."[35] For instance, in the case of copper, its vitriol, its burnt brass, its aes crustum (also Crocus of Copper), and its verdigris, were all mortified copper, for the production of which, he offered several recipes in Book V of *Concerning the Nature of Things.* [36]

Burnt Copper[37]

The mortification of the metals, then, transformed them into useful medicinal substances. According to Paracelsus, the Planets possessed unique spirits (properties) that were imparted to their corresponding metals. He believed the metals in turn could be transformed into a form that would allow them to pass on their properties (or spirits) as cures or protections for the human body. Paracelsus referred to the Spirit of Venus as a red spirit that was produced from "... a thick elemental mixture of ... [the White Spirit or Tincture of Luna (silver)] ... to which also it is subject...." Venus, then, contained the same elemental mixture as Luna only in a thickened version. Presumably, the elemental mixture referred to the principles of Mercury, Sulfur, and Salt. The most rarefied mixture was seen in Sol (gold). Luna (silver) would then be a little denser, and Venus (copper) would be even more so.

Brass Cremator[38]

This red spirit of Venus (copper), according to Paracelsus, was more perfect than the spirits of the succeeding metals (i.e., tin, iron, lead), and so remained more constantly and fixedly in the fire than the rest. Compared to Mars (iron), Venus was more resistant to air and moisture, less dense, and combustible. When added to other metals, it embrittled them, causing them to lose their malleability.[39] In a similar fashion, if the spirit of Venus, as conveyed through some transformation of copper, was used to treat the human body, it would protect wounds from air and moisture. If, however, the Spirit of Venus was taken for a disease for which it was not suited, it would produce contraction (embrittlement) of the limbs.[40]

Like other alchemists before him, Paracelsus was well acquainted with verdigris as a common and popular ingredient for coloring agents and pharmaceuticals. An example of one of Paracelsus's recipes was given in *Concerning the Nature of Things*:

> *The verdigris used in medicine admits of the ensuring process: Take plates of copper and smear them with the following compound: Take equal quantities of honey and vinegar, with a sufficient quantity of salt to make the three together the consistency of thick paste. Mix thoroughly, and afterwards put in a reverberatory, or in a potter's furnace, for the same time as the potter bakes his vessels, and you will see*

a black substance adhering to the plates. Do not let this circumstance cause you anxiety or detain you at all; for if you suspend or expose those plates in the open air, in a few days the substance will turn green and will become excellent verdigris, which may be called the balsam of copper, and is highly esteemed by all physicians.[41]

Through the concepts of generation and mortification, spirit and body, and the Sun and Planets, Paracelsus personified and animated the metals, including copper as an earthly embodiment of Venus.

Georgius Agricola (24 March 1494 – 21 November 1555)
Though Agricola acquired much of his technical nomenclature from Pliny, he offered clearer operational terms and descriptions of copper and its associated compounds. For instance, he described Regulare or Yellow Copper as a yellowish-red copper produced in smelters that separated silver from copper, while Caldarium Copper was a dark yellowish-red copper, and differed from Regulare in that it was easier to cast than forge. Agricola discussed a third alloy of copper, oreichalkos, which contained zinc and had a golden color; Pliny did not offer such detail.

In his discourse on assaying ores, Agricola named four forms of copper as re-agents—copper filings (Aeris scobs elimata), Copper Scales (Aeris squamae), Copper Flowers (Aeris flos), and Roasted Copper (Aes ustum, a term most likely acquired from Pliny, which was probably an artificial sulfide).[42] These four forms were included in Agricola's list of fluxes, or his term "Additions," which were used for reducing, oxidizing, sulfurizing, desulfurizing, and collecting agents during either assaying or smelting operations. Most likely, copper filings were finely divided copper metal, while the others were probably all cupric oxide. The scales resulted from hammering the hot metal. The flowers (flos), on the other hand, came off spontaneously when the hot copper bars were quenched in water. Flowers of Copper, also called anthos chalkos by the Greeks, consisted of small particles of copper that were loosened from the main body of copper and appeared as millet. Finally, the Roasted Copper was produced from the calcination of copper. In addition, Agricola pointed out that the particles of the Flower of Copper were finer than the Copper Scales—also referred to as crematum Copper. Flowers of Copper and Copper Scale had the same properties as Roasted Copper—a certain acridness with an astringent quality. The Flowers, however, were more tenuous than either Copper Scale or Roasted Copper, and for that reason, according to Agricola, were mixed with certain remedies to cure irritations and ulcers of the eyes.[43]

Aes Ustum (Copper or Brass Cremator)[44]

Agricola also pointed out that copper often contained imperfections, especially when exposed to acid solutions, and as such, was referred to as aerugo (verdigris). He identified two types of verdigris—one was ". . . smooth and another was full of holes. . . ." The smooth variety, made from copper and vinegar, was called santerna and was used for soldering gold.[45]

Martinus Rulandus (1532-1602)
Rulandus's lexicon offered a few additional terms related to copper; these included Cuprum, Cancer, Cuprum Rebeum Incompletum (Red Copper), and Mupur, but he did not provide specific descriptions or terms that would have further defined the meaning of these terms. In regard to copper compounds, he did offer another term for verdigris, which was Flor Aeris, and, for Burnt Copper the terms Arcos, Aycophes, and Azaphora.[46] Unfortunately, he did not offer any clues as to the exact meaning or the origins of these terms.

Nicholas Lemery (17 November 1645 – 19 June 1715)
Where others took the attribution of Venus to copper for granted, Lemery gave a reason for the connection: "It is called Venus, because this Planet was thought to govern it particularly, and bestow its Influence upon it, and for this Reason there hath been attributed unto it the Virtue of increasing seed and curing the Diseases of those Parts that serve for Generation." Since copper had a corrosive quality, he advised that it not be consumed orally: "But because Copper contains in it a Corrosive Quality, I would advise no Body to use it inwardly." In addition to the association of Venus and copper, Lemery affirmed other common terms for copper used in his day—Rose Copper, Aes, and Cuprum. Refined

Copper, known as Rose Copper, was obtained by the melting and remelting of copper several times: "To refine it, they melt it two or three Times, for upon every Fusion, some of the gross earthy Parts are separated from it, and then it is called Rose Copper."[47]

Since iatrochymistry was Lemery's primary interest, his focus on copper was directed toward medicinal uses, the preparation of which involved the calcination of copper to produce Crocus of Copper or copper oxide:

> *To Calcine Copper is to purify it from its more volatile parts by means of common sulphur and fire in order to render it the more compact. Stratify plates of copper with powdered sulphur in a large crucible, cover the crucible with a cover that hath a hole in the middle to give vapors vent. Place your crucible in a wind furnace and light a very strong fire about it until there rise no more vapors, then draw off your plates as they are hot and separate them; this is the Aes crustum that is used in outward remedies to deterge. It is powdered in a mortar.*
>
> *Purification of calcined Copper—the recipe produces Crocus of Copper. This second perfection of copper is to render it fair to the eye and of a high color. Take what quantity you please of calcined copper, as above, heat it red-hot in a crucible placed among burning coals, and cast it red-hot into a pot wherein you shall have put enough oil of linseed to swim above it four fingers; cover the pot presently, for otherwise the oil would take fire; let the copper steep till the oil is grown pretty cool, separate it, and put it to heat again in the crucible, then cast it into oil of linseed; continue to make it red-hot and quench it in the oil some several times. You must change your oil every third time, and you'll have a copper well purified and of its former color. If you calcine it once again to consume the oil and powder it, you'll have a crocus of copper that is deterrent and good to eat the proud flesh of wounds and ulcers.*[48]

As pointed out by Agricola, the Aes ustum was Roasted Copper, a scale of copper oxide. Sulfur was added to the crucible, according to Lemery, to cleanse the copper of its superficial sulfur, and the resulting Aes ustum, then, was used as the basis for making Crocus of Copper. The purpose for the use of linseed oil is not clear. Perhaps Lemery used it for its drying properties. Linseed oil oxidizes rapidly and may have been useful for the further purification of the calcined copper. A hint may have been given by the section on "Purification of Calcined Copper," in which he stated that the operation was "... to render it fair to the Eye, and of a high Color." The implication was that the calcined copper, initially dull red, would be refined to yield a brighter color, one that would be more pleasing to the eye. The dullness of color may have indicated the presence of remaining impurities. Whatever the actual purpose of the linseed oil, Lemery referred to the result as Crocus of Copper and used it to treat the excess scabbing of wounds and ulcers.

Another concoction was Vitriol of Copper or Vitriol of Venus, which was also used to reduce the superfluous flesh from scabbing: "They (the crystals of Vitriol of Copper) are caustic, and are used to consume superfluous or proud Flesh."

> *Vitriol of Copper or Venus—*
> *This Operation is a Copper opened, and transformed into a Vitriol by Spirit of Nitre. Dissolve two Ounces of Copper cut into little Pieces, in five or six Ounces of Spirit of Nitre, pour the Dissolution into a Glass Cubicrit, and evaporate in Sand about the fourth Part of the Liquor, put that which remains into a Cellar, or some other cool Place, and let it lie there five or six Hours, you'll find Blue Crystals, separate them, and continue to evaporate and crystallize, till you have drawn them all, dry these Crystals, and keep them in Vial well stopped. They are caustic, and are used to consume superfluous or proud Flesh.*[49]

"Copper opened . . ." referred to the dissolution of copper by the Spirit of Nitre—nitric acid. The solution was allowed to evaporate off, leaving a precipitate of blue crystals, which would be indicative of copper nitrate:

> *The great effervescence that happens does proceed from the suitable pores of copper to the edges of spirit of nitre, so that they can make their entrance and jostle with a good force, for when these edges, which did before swim with all liberty in a liquid, do find their motion checked in the body of the metal, they do strive to disengage themselves and do thereby separate the parts of the copper. It is this violent separation which causes the effervescence and heat, for the acid edges striking strongly against the solid parts of copper do cause a great agitation in the liquor and by that means do excite a heat, much after the manner*

as when two sound bodies are beaten against one another violently, they grow so hot as even to strike fire." [50]

In addition to describing the products of the copper-nitric acid reaction, he also offered an explanation of the dynamics involved. According to Lemery's Acid-Alkali Theory, the metallic copper exhibited suitable pores that were receptive to the points (edges) of the acid particles. Once the acid had penetrated the metal, the edges of the acid corpuscles became restricted and because of their nature sought to disengage themselves, causing the copper parts to separate (dissolve). Since the effort of the acid particles to disengage themselves was great, the action generated great agitation and heat.

Another remedy put forth by Lemery was Spirit of Venus. The Spirit of Venus was described as an acid liquor prepared from the Crystals of Venus, which were verdigris (a copper salt of acetic acid), prepared by treating copper with vinegar:

> *The Crystals of Venus are the Particles of Copper impregnated with the Spirits of Vinegar, and reduced into a Form like Salt of Vitriol. . . . These crystals may also be called Verdigrease, because it is prepared with distilled Vinegar.* [51]

After the Crystals of Venus were prepared, the Spirit of Venus could be produced and used to treat epilepsy, palsy, apoplexy, and ". . . other diseases of the Head":

> *The Spirit of Venus is an acid Liquor, drawn from the Crystals of Venus by distillation. Put what Quantity you please of the Crystals of Venus prepared with distilled Vinegar . . . into a Glass Retort, whose third part remains empty. Place your Retort in Sand, and fitting it to a large Receiver, and luting well the Junctures, give a small Fire at first, to drive out a little insipid Phlegm, this Phlegm will be followed by a volatile Spirit. Then augment the Fire by Degrees, and the Receiver will fill with white Clouds. Towards the latter End kindle Coals round about the Retort, that the last Spirits may come forth, for they are the strongest. When you see the Clouds disappear, and the Recipient grow cool, put out the Fire, unlute the Junctures, and pour all that which is in the Recipient into a Glass Body to distil it in Sand until it is dry. This is the rectified Spirit of Venus.* [52]

CHAPTER 7

Like Paracelsus, Lemery's major concern was iatrochymistry, experimenting with various substances to develop medical treatments for his patients. The application of copper to that end was no exception.

Conclusion

From the mists of antiquity into the pre-modern age of the 17th and 18th centuries, naturalists, metallurgists, and alchemists alike have been enraptured by the Spirit of Venus embodied in her earthly counterpart—copper. Building on the work of those who came before, they have sought to express the beauty, to uncover the nature, and to exploit the properties of copper. Indeed, as Venus was highly regarded in the planetary pantheon alongside Sol and Luna, so was copper elevated along with gold and silver. As Paul of Taranto (Pseudo-Geber) asserted, as Venus was the medium between Sol and Luna; so, too, was copper the medium between gold and silver. Copper's beauty and utility placed it between the two noble metals. It was believed that Venus's metallic counterpart could be transmuted into either gold or silver, and, if purified or alloyed, could serve as practical tools, implements, and weapons, or serve as beautiful ornaments, pigments, and dyes. Just as important, though, copper compounds were discovered to convey the spirit of Venus to the human body and provide cures and treatments to restore the body's health and beauty.

Copper or Venus[53]

Notes

1. Michael Stapleton, *A Dictionary of Greek and Roman Mythology*, (New York: Bell Publishing Co., 1978), 29. Also, Hesiod, *Theogony*, Lines 173-206. Hesiod, *Theogony*, line 173 (tufts.edu). Retrieved 10-Nov-2023. See also Plato's *Symposium*.
2. Philip Wheeler, *Alchemical Symbols, Fourth Edition, R.A.M.S. Library of Alchemy, Vol. 21,* (Kansas City, MO: R.A.M.S. Publishing Co., 2018), 32.
3. John Emsley, *Nature's Building Blocks: An A-Z Guide to the Elements*, (Oxford: Oxford University Press, 2011), 147.
4. Copper, *Merriam-Webster Dictionary*. 2018. Retrieved 4-Nov-2023.
5. T. A. Rickard, (1932), "The Nomenclature of Copper and its Alloys," *Journal of the Royal Anthropological Institute*. 62: 281–290.
6. Catherine E. Housecroft & Alan G. Sharpe, *Inorganic Chemistry*, 5th edition, (Harlow, UK: Pearson, 2018), 789.
7. Pliny the Elder, *Natural History*, Book XXXIV Chapter 29. https://www.perseus.tufts.edu/. Retrieved 10-Nov-2023. See also the chapter "The Vitriols."
8. Pliny the Elder, *Natural History*, Book XXXIV, Chapter 2. https://www.perseus.tufts.edu/. Retrieved 10-Nov-2023.
9. Wheeler, 26.
10. Pliny the Elder, *Natural History*, Book XXXIV, Chapter 20. https://www.perseus.tufts.edu/. Retrieved 10-Nov-2023.
11. Pliny the Elder, *Natural History*, Book XXXIV, Chapter 23. https://www.perseus.tufts.edu/. Retrieved 10-Nov-2023.
12. Pliny the Elder, Natural History, Book XXXIV, Chapters 24 and 25. https://www.perseus.tufts.edu/. Retrieved 10-Nov-2023.
13. Georgius Agricola, *De Re Metallica*, translated from the first Latin edition of 1556 by Herbert Clark Hoover and Lou Henry Hoover, (New York: Dover Publications, Inc., 1950), 405. See also, Pliny the Elder, *Natural History*, Book XXXIV, Chapter 2. https://www.perseus.tufts.edu/. Retrieved 10-Nov-2023.
14. Karin C. Ryding (1994) "Islamic Alchemy According to Al-Khwarizmi," *Ambix*, 41:3, 121-134.
15. Richardson, H. Wayne (2000). "Copper Compounds". *Ullmann's Encyclopedia of Industrial Chemistry*.
16. Wheeler, 15, 4, 82.

CHAPTER 7

17. J. R. Partington (1937), "Albertus Magnus on Alchemy," *Ambix*, 1:1, 3-20.
18. E.J. Holmyard, *The Works of Geber*, (Kessinger Publishing, 1928), 68.
19. Th. Rehren, "Crucibles as reaction vessels in ancient metallurgy," in P.T. Craddock and J. Lang (eds), *Mining and Metal Production through the Ages*, (London. The British Museum Press, 2003), 207-215. See also, J. Bayley, "Medieval precious metal refining: archaeology and contemporary texts compared," in Martinón-Torres, M and Rehren, Th (eds) *Archaeology, history and science: integrating approaches to ancient materials* (Left Coast Press, 2008), 131-150.
20. Cementation—Acting upon a substance by mixing it in layers with a powdered (often corrosive) material, such as lime. This mixture is then made to react and weld together by heating to a high temperature in a cementing furnace. https://alchemywebsite.com/alch-pro.html. Retrieved 12-Nov-2023.
21. Holmyard, 184.
22. William R. Newman, *The Summa Perfectionis of Pseudo-Geber: A Critical Edition, Translation, and Study*, (Leiden: E.J. Brill, 1991), 676n74, n75.
23. Dried or darkened as if by heat; scorched; burnt. *Webster's Unabridged Dictionary of the English Language*, (New York: Random House, 2001).
24. Holmyard, 105-107.
25. Newman's translation was helpful in understanding this excerpt from Holmyard's *The Works of Geber*. See Newman, 707-708.
26. Newman, 676, 676n76.
27. Holmyard, 135-136.
28. Ibid, 155.
29. Ibid, 215.
30. Paracelsus, *The Hermetic and Alchemical Writings of Aureolis Philippus Theophrastus Bombast, of Hohenheim, Called Paracelsus the Great*, Edited by Arthur Edward Waite in two volumes, (Mansfield Centre, CT: Martino Publishing, 2009), Vol. II, 354.
31. In Paracelsus's *A Book about Minerals*, he defined cachimiae as imperfect bodies and considered it a genus of minerals to which belonged such substances as marchasites, pyrites, genera of antimony, varieties of arsenicalia, talcs, auripigments, and "... many cachimiae of this kind ..." Paracelsus, Vol. I, 255-256. Also, the dross of metals, Paracelsus, Vol.

II, 358. Also, according to Rulandus, cachimiae is white argentiferous chalk, Rulandus, 77.
32. Paracelsus, Vol. I, 7n. See also *De Elemento Aquae*, Lib. IV., Tract III., c.1, c.4. and Martinus Rulandus, *A Lexicon of Alchemy, Containing a Full and Plain Explanation of All Obscure Words, Hermetic Subjects, and Arcane Phrases of Paracelsus*, (Zachariah Palthenus, bookseller, in the free republic of Frankfurt, 1612), 36-37.
33. Paracelsus, Vol. I, 256.
34. Andrew Sparling (2020), "Paracelsus, a Transmutational Alchemist," *Ambix*, 67:1, 62-87.
35. Paracelsus, Vol. I, 139.
36. Ibid, Vol. I, 138-194.
37. Wheeler, 32.
38. Wheeler, 15.
39. Paracelsus, Vol. I, 76-78.
40. Ibid.
41. Ibid., Vol. I, 141.
42. Agricola, 232, 232n7, 233n9, 538.
43. Georgius Agricola, *De Natura Fossilium (A Textbook of Mineralogy)*, Translated from the first Latin Edition of 1546 by Mark Chance Bandy and Jean A. Bandy (Mineola, NY: Dover Publications, Inc., 2004), 176, 177, 194-197.
44. Wheeler, 15.
45. Agricola, De Natura Fossilium, 176, 177, 194-197.
46. Martinus Rulandus, *A Lexicon of Alchemy, Containing a Full and Plain Explanation of All Obscure Words, Hermetic Subjects, and Arcane Phrases of Paracelsus*, (Zachariah Palthenus, bookseller, in the free republic of Frankfurt, 1612), 37, 118, 120, 147, 322.
47. Nicolas Lemery, *A Course of Chymistry, The Fourth Edition*, London, 1720. (Published by Creative Media Partners on behalf of Gale Research, Inc.), 101-102.
48. Ibid., 104-105.
49. Ibid., 105.
50. Ibid., 105-106.
51. Ibid., 106.
52. Ibid., 107.
53. Wheeler, 32.

Chapter 8

The Masks of Antimony

Antimony

Many an alchemist sought to conceal the true meaning of their writings with words or phrases that served as masks. These "*decknamen*"— German, singular, "*deckname*," (i.e., a cover name, secret name, or alias)—functioned as an enigmatic figure or image within an allegorical writing. *Decknamen* were used along with other techniques of concealment and revelation, which included parathesis,[1] dispersion,[2] and syncope.[3] Alchemists used these techniques of an extended conceit to conceal experimental knowledge, and yet communicate, discreetly, to those who have the knowledge to decipher the texts. The purpose for this was to protect "trade secrets" and allay apprehensions of economic disruption should the Philosopher's Stone become common knowledge. Yet a means was also provided to those dedicated to decoding the writings. Another reason, offered by Newman and Principe in their work, *Alchemy Tried in the Fire*, involved intellectual entertainment. For an early modern mind, creating and solving riddles and allegories were a favored pastime, but also as a quest for knowledge. Heraclitus's assertion that "nature loves to hide" was taken seriously. Often alchemical writings, when juxtaposed with the author's other works, could give up the meaning of the employed *decknamen* to dedicated and select readers.[4]

One other note a prospective alchemical sleuth needs to keep in mind is that of context. Since different alchemists would use the same word, phrase, or image to convey different meanings, or attach the same meaning to different words, phrases, or images, an interpretation cannot stray too far from an author. It may be possible to extend an interpretation to other authors who are closely tied together. For instance, Starkey expounded on the writings of George Ripley, under the pseudonym of Eirenaeus Philalethes, in an attempt to reproduce Ripley's observations and findings. In that case, the meanings uncovered by Starkey would apply to Ripley's *decknamen*. Also, in much of the literature, the illustrations were added after the text and were meant to illustrate what was in the text. Deciphering these alchemical emblemata, then, was dependent on the text.[5]

Antimony was a primary candidate for concealment since it was instrumental in the formation of various arcanum. Alexander von Suchten (German physician, 1525-1570)[6] believed a potent medicine could be made from antimony by separating the regulus of antimony from its ore, stibnite, then alloying the regulus with silver and using the resultant alloy to actuate common quicksilver. The goal was to isolate the "volatile gold" within the regulus, which was eventually turned into "potable gold" or "Philosophical Gold"—the desired medicinal arcanum. The actuated quicksilver, it was thought, also had the power to penetrate the metals and separate their Mercury and Sulphur from one another. Suchten's antimonial mercury, then, was a crucial desideratum since he believed that the process for making the Philosopher's Stone started with the dissolution of gold into its principles.[7]

Following Suchten, George Starkey developed a recipe for making an amalgam of mercury, antimony, and silver. He had hoped that the product would be a "philosophical mercury" in which gold would melt and undergo a series of changes leading to the transmutational agent—the Philosopher's Stone or the Elixir. He employed a metallurgical refining process using antimony to derive "sophic mercury," which would lead to the Philosopher's Stone. Instead of using antimony to purify gold, however, Starkey applied it to quicksilver in the hopes that the metallic antimony would purge the mercury, making it so subtle and penetrative that it would not only amalgamate to gold but dissolve the precious metal into its primal ingredients, which could recombine with the mercury to make the Philosopher's Stone.[8] Starkey considered this knowledge a secret of "highest value," to be protected. Starkey regarded the importance of protecting this secret so highly that he refused to reveal it to Benjamin Worsley (English physician, 1618-1673) when the latter offered to create a joint venture which would enable Starkey to pay off his debts. Starkey did, however, reveal his secret, but only to Robert Boyle in a letter written in the spring of 1651.[9]

These examples of antimony's role in the production of various arcanum illustrate its importance to the alchemical arts and the need for a protective subterfuge such as the use of *decknamen*.

One of the earliest writings in which this masking technique may have appeared was in the works of Pseudo-Democritus. It is unclear whether the author meant to disguise the antimony ore using the word "lead" or not, since the ore, stibnite, and lead shared similar properties,[10] and the author could have only meant a simple substitution of words.

> *Process silver pyrite that is also called siderites as is customary, so that it may be melted. It will melt with nitre (?), or white litharge, or Italian stibnite. So reduce it to powder, mixing with lead [lit. sprinkle it on lead]: I do not mean 'lead' in its general use—do not be misled—but with our black [lead] composed from Coptic stibnite and litharge, or according to your knowledge. So roast it and, when it turns yellow, lay it on the matter: you will dye it. For nature delights in nature.*[11]

A more enigmatic image used for stibnite was that of the wolf. What, then, did the wolf conceal, and did the true identity of the wolf go by other aliases, establishing the wolf as a member of a family of secret names? An excerpt from Basil Valentine's book *Of the Great Stone of the Ancients* presents an excellent starting point for the required detective work.[12]

> *... the king's crown should be pure gold, and a chaste bride should be married to him. Take the ravenous grey wolf that on account of his name is subjected to bellicose Mars, but by birth is a child of old Saturn, and that lives in the valleys and mountains of the world and is possessed of great hunger. Throw the king's body before him that he may have his nourishment from it. And when he has devoured the king, then make a great fire and throw the wolf into it so that he burns up entirely; thus will the king be redeemed. If this is done thrice, then the lion has conquered the wolf, and nothing more to eat will be found on him; thus is our body completed at the start of our work.*[13]

From *The Alchemy Collection: The Works of Basil Valentine*[14]

Fortunately, Principe has solved this riddle and has shared it in his book *The Secrets of Alchemy*. Valentine describes a purification process. Given that he is writing about metallic transmutation, the king is the king of metals, gold, and whose body is fed to a "ravenous grey wolf." So, what is this devouring wolf? Valentine offers two hints. First, the wolf is subjected to Mars, the Roman god of war, who is associated with iron in the nomenclature of metals and weapons of war, including a spear. Next, the wolf is said to be "... a child of old Saturn...." Saturn in planetary nomenclature is associated with lead, and a child of Saturn would point to something closely associated with lead. The substance associated with both of these attributes of the wolf was stibnite, an ore of antimony (antimony trisulphide). Stibnite was widely held to be related to lead and was used to purify gold. Also, the German name (Valentine's native tongue) for stibnite was Spiessglanz—"spear-shine"—a reference to the needle-like crystalline structure of the ore.[15]

The wolf, however, was one among many of antimony's "*decknamen* family." Another was the Daughter of Saturn, also called Saturnia, described in Philalethes' commentary on Ripley[16]:

> *Take then the most beloved Daughter of Saturn, whose Arms are a Circle of Argent, and on it a Sable Cross on a Black Field, which is the signal note of the great world, espouse her to the most warlike*

God, who dwells in the house of Aries, and thou shalt find the Salt of Nature, with this Salt acuate thy water, as thou best knowest, and thou shalt have the Lunary bath in which the Sun will be amended.[17]

Here we see the Daughter of Saturn's arms forming a circle on top of which is a cross, the symbol for antimony. She was then to be espoused to the warlike god, an allusion to Ares or Mars, which was a *deckname* for iron. What then is meant by "dwells in the house of Aries?" In Ptolemaic astrology, one of the two celestial houses, in which the planet Mars resides, was Aries, the Ram, which, along with Leo and Sagittarius, constitutes a trinity of constellations known as the fiery triplicity. Thus, the phrase alluded to the subjection of antimony to a fiery heat. What was the marriage of crude antimony, Saturn's daughter, and iron, the god of war? In Chapter V of Philalethes' *Introitus*[18], a reference to the marriage of antimony and iron led to a product which had a center that was ". . . astral, irradiating the earth all the way up to its surface with its brightness" (Here is an example of dispersion). The product of the marriage was a substance with a star-like appearance, which points to regulus antimony (metallic antimony). The star-like formation occurred when the molten antimony cooled slowly under a covering of slag. The crude antimony (stibnite) was reduced, using iron, to form the star regulus antimony, the Salt of Nature, which, in turn, was mixed with mercury.[19]

Star Regulus of Antimony from Chymical Products: The Chymistry of Isaac Newton Project (indiana.edu)

Philalethes reveals additional cover names for antimony. He also tells us that antimony is known to him as "our Chaos," the product of the marriage between Saturnia and Mars, in the *Introitus*. With this knowledge, we have a key for many antimony aliases. In Chapter II of the *Introitus,* the following is found:

ALCHEMICAL MATTERS

This Chaos is called our Arsenic, our air, our Luna, our Magnes, our Chalybs, but in diverse respects, because our matter undergoes various states before our Regal Diadem is extracted from the menstrual blood of our whore. So learn who the comrades of Cadmus are, and who the Serpent who ate them, [and] what the hollow oak, on which Cadmus transfixed the Serpent. Learn what the Doves of Diana are, which conquer the Lion by beating him, the green Lion, I say, which is really the Babylonian Dragon, killing all by means of his venom.[20]

Once particular images and symbols for antimony have been identified, the presence of antimony could be seen in alchemical emblemata. Again, to understand the full meaning of particular emblemata, care must be taken in regard to related texts and historical context. Examples of references to antimony in emblemata include the image of the wolf devouring the king in Michael Maier's work *Scrutinium chymicum* and an illustration from the 1695 *Opera omnia* of Philalethes in which three symbols of antimony are shown. These symbols, read from right to left, show a cross within the antimony symbols acting as flasks, being decomposed and replaced with a growing star, which becomes the star regulus, the symbol for metallic antimony. The rendered process is the purgation of stibnite as it loses its sulfur to become metallic antimony. Within this illustration are additional images of antimony. According to Starkey's (the true author of Philalethes's works) *The Marrow of Alchemy*, the urinating cherub-like figure in the upper right is the "son of Saturn," a reference used by Starkey to symbolize crude antimony. Just as the son of Saturn is crude antimony, "old Saturn's pisse" is the star regulus. The crude antimony washed with urine alludes to the blackness and "great stink" given off when the metallic antimony is amalgamated with silver and quicksilver.[21]

Purgatio materiae et reduction geniti crudi in Genitorem coctum, ut urina sua lavet mercurium (from *Gehennical Fire*, fig. 3b, 164-165).

CHAPTER 8

From *Scrutinium chymicum*

Decanamen were not the only means of encoding terminology. Alchemists employed a variety of graphic symbols for their substances, processes, and apparatus, which varied in meaning and form among different alchemists and over time. The following are a few symbols used for antimony:

First figure: Antimony tri-sulphide or Stibnite (Sb_2S_3)—often simply referred to as Antimony. Second figure: Regulus of Antimony (metallic antimony). Third figure: Star regulus of Antimony (crystalline Sb) or sometimes used for sal ammoniac (NH_4Cl).

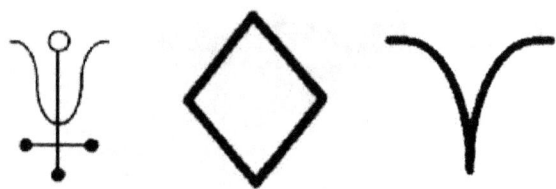

Additional symbols for antinomy [22]

Outside of the immediate family of antimony, *deckname*s were the related cover names for the associated compounds of antimony. For the sake of brevity, these are listed in the table below:

Alchemical Names for Antimony Compounds[23]	
Alchemical Name	Meaning
Butter of antimony	Antimony tri-chloride, $SbCl_3$
Tartar Emetic	Antimony Potassium Tartrate, $K_2Sb_2(C_4H_2O_6)_2$
Crocus of antimony	Brownish-yellow impure sulphide of antimony and sodium, formed as a scoria or slag in the smelting of antimony
Flowers of antimony	Crystals of the trioxide formed when the metal is sublimed
Antimony blend*e*	Kermesite, an antimony oxysulphide (Sb_2OS_2) called red antimony
Lupus metallorum	The grey wolf or stibnite, used to purify gold, as the sulphur in the antimony sulphide bonds to the metals alloyed with the gold, and these form a slag which can be removed
Glass of antimony	Vitreous oxide of antimony mixed with sulphide
Saffron of antimony	Golden sulphide of antimony, the pentasulphide of antimony (Sb_2S_5)
Antimony vermilion	Red or orange colloidal antimony sulphide precipitate

Alchemical Names for Antimony Compounds[23]	
Red Flowers of Antimony	Antimony sulphide
Powder of Algaroth	White powder of Antimonious Oxychloride (SbOCl)
Kermes Mineral	Soft brown-red powder consisting essentially of oxides and sulphides of antimony
Liver of Antimony	Fused antimony sulphide
Antimony glance	Gray antimony, another name for a form of stibnite
Antimony ochre	Stibiconite, an Antimony mineral, $SbO_3(OH)_2$
Plimmer's Salt	Sodium Antimony Tartrate, $Na(SbO)C_4H_4O_6$
Antimony bloom	Antimony trioxide (Sb_2O_3) which appears as the mineral Valentinite
Vinegar of antimony	Supposed acetate of antimony mentioned in texts of Basil Valentine. A triacetate of antimony exists but it is unstable being easily hydrolyzed
Antimony black	Metallic antimony

Conclusion

Many alchemists believed that the valuable nature of such substances as the Philosopher's Stone, the Elixir, Philosophical Mercury, and Philosophical Gold necessitated that the ingredients, processes, and apparatus involved in the creation of these arcanum must be hidden from the unscrupulous and undeserving. Thus, elaborate means (*decknamen*) were devised to conceal the true identity of the substances and processes involved. Antimony was one such ingredient among many that received this treatment. Various masks were created for antimony: wolf, Daughter of Saturn, Saturnia, lead, Chaos, Arsenic, Luna, air, Chalybs, Magnes, son of Saturn, and even the vulgar "old Saturn's pisse." In addition, diverse symbols and images were added to the mix. Even the related compounds of antimony were concealed within aliases. All designed to confound those who lacked the scruples or the diligence required of true adepts, but to hold promises of great knowledge and riches to those who were true to the Art.

Notes

1. Parathesis—The placing of grammatically parallel words or phrases together; apposition; and in the case of alchemical writings involving the needless multiplication of processes or ingredients.
2. Dispersion—Scattering pieces of a single item through a text or set of texts, which could be used as hints or keys to the meaning of a word or image.
3. Syncope—The abbreviation of a process, typically omitting one or more steps or ingredients.
4. William R. Newman and Lawrence M. Principe, *Alchemy Tried in the Fire: Starkey, Boyle, and the Fate of Helmontian Chymistry*, (Chicago, University of Chicago Press: 2002), 180-187.
5. Lawrence M. Principe, *The Secrets of Alchemy*, (Chicago, University of Chicago Press: 2013), 74.
6. Alexander von Suchten was a German physician who, for his own alchemical work, relied heavily on Paracelsus's work.
7. Newman and Principe, 50, 108. For further discussion, see 108n44.
8. William R. Newman, *Gehennical Fire: The Lives of George Starkey, an American Alchemist in the Scientific Revolution, (Chicago,* The University of Chicago Press: 1994), Forward.
9. Ibid., 67-69.
10. The name, Pseudo-Democritus, refers to anonymous authors of a number of Greek writings that were attributed to Democritus (c. 460–370 BC). Greek alchemical writings attributed to Democritus, whose author has sometimes been identified as Bolos of Mendes. More recently, however, the authorship is attributed to an anonymous author active during the second half of the first century AD, c. 54–68 AD. These writings are some of the oldest alchemical works in existence. In their original form, they probably consisted of a series of four books on dying: two books on dying metals gold and silver, one on dying stones, and one on dying wool purple. See Matteo Martelli, *The Four Books of Pseudo-Democritus*: Sources of Alchemy and Chemistry, Sir Robert Mond Studies in the History of Early Chemistry, *Ambix* Volume 60, Supplement 1, 2013 (London and New York, Rutledge: 2017).
11. Litharge is usually understood as lead monoxide, PbO, which presents two allotropic forms in different colors, yellow and red. Ancient sources, for example, Pliny and Dioscorides, referred to different methods

for purifying and whitening litharge: one was to process the substance with salt and sodium carbonate, and the other was to wash with vinegar to produce white lead acetates. Coptic stibnite was presumably a lead-antimony alloy. Nitheos could be a cover name. See Martelli, S87-S89, S155, S217-S219 n 27, 28, 29, and 30, S252-S253 n 7.
12. Likely a pseudonym used by one or several 16th-century German authors. *Of the Great Stone of the Ancients* was possibly published in 1599.
13. Principe, 144-145.
14. Basil Valentine, *The Alchemy Collection: The Works of Basil Valentine: Including The Triumphant Chariot of Antimony, The Twelve Keys, and Of Natural & Supernatural Things,* 3rd edition, Adam Goldsmith, editor (Vitriol Publishing: 2013).
15. Principe, 146.
16. Philalethes, *Expositions Upon Sir George Ripley's Compound of Alchymy, &c.,* comments on Stave XI, in *Ripley Revived*.
17. Newman, *Gehennical Fire*, 127.
18. First published as *Introitus apertus ad occlusum regis palatium*, Amsterdam, 1667 and a few years later issued in an English edition *Secrets reveal'd: or, an open entrance to the shut-palace of the king*, London 1669. The Alchemy Website (levity.com).
19. Newman, *Gehennical Fire*, 128-129.
20. Ibid 129-131.
21. Ibid, 164-165.
22. Philip Wheeler, *Alchemical Symbols, 4th edition: RAMS Library of Alchemy Vol. 21*, (Kansas City, MO, RAMS Publishing Company, 2018), 19.
23. Table information obtained from The Alchemy Website (https://www.alchemywebsite.com).

Chapter 9

Arsenic and Old Alchemy

Introduction

Arsenic[1] has been known since the Bronze Age (3300-1200 BC) when it was used to make bronze[2] harder. The alloy was known as arsenical-bronze in which arsenic, as opposed to or in addition to, tin or other constituent metals, was combined with copper to make bronze. The use of arsenic with copper or iron, either as the secondary constituent or with another component such as tin, resulted in a stronger final product with better casting behavior.[3] Arsenic is present in a number of copper-containing ores[4] and therefore some contamination of the copper with arsenic would have been unavoidable. The extent to which arsenic was intentionally added to copper and the extent to which its use came about simply from its natural presence in copper ores treated by smelting to produce the metal, however, has not been clear, but the adverse health effects of the mining and alloying of copper were certainly known at this time. Arsenical oxide was readily lost from arsenical copper as an extremely toxic white fume during the process of hot forging, and thus the toxicity of arsenical materials would have been certainly known by early copper miners and smelters.[5]

Arsenic[6]

CHAPTER 9

The early understanding of arsenical chemistry was primarily driven by the sulfide ores of arsenic, orpiment and realgar, and their metallurgical, tinging (coloring), and iatrochemical applications. The progression of this understanding, however, was anything but linear—more like eddies in a stream that swirl around and circle back, and yet, somehow contribute to the progression of the stream. Even though all of these applications were present to some extent throughout antiquity and the Middle Ages, the iatrochemical uses did not significantly gain ascendancy until about the 15th and 16th centuries. From antiquity to the 15th century, the focus was on the metallurgical (transmutation is included as a metallurgical operation), the tingeing, and the dyeing applications of arsenic. Then, from the 15th century on, especially with the polemics of Paracelsus, the iatrochemical (medicinal) applications ascended. An initial comparison of the alchemical writings may reveal seemingly discordant descriptions and terminology; however, a more in-depth look, utilizing modern terminology, may elucidate the similarities and provide a sense of the progression of arsenical chemistry.

Orpiment and Realgar

The two arsenic ores most often mentioned by the alchemists were the arsenic sulfides, orpiment (As_2S_3) and realgar (As_4S_4 or AsS).[7] Orpiment and realgar are closely related minerals often categorized in the same group. As arsenic sulfides, they belong to the monoclinic crystal system. They are often found in the same deposits and can form in the same geologic environments. As a result, orpiment and realgar share similar physical properties and histories of use by humans. Orpiment was a deep-colored, orange-yellow to golden-yellow mineral found in volcanic fumaroles, low-temperature hydrothermal veins, and hot springs. It was formed either by sublimation or as a byproduct of the decay of realgar.[8] (After a long period of exposure to light, realgar changes form to a yellow powder known as pararealgar (β-As_4S_4) or arsenolite and orpiment. It was once thought that this powder was the yellow sulfide orpiment, but it is actually a distinct chemical compound.)[9]

Alchemical Symbol for Orpiment[10]

ALCHEMICAL MATTERS

Orpiment, derived from the Latin *auripigmentum* (*aurum*, "gold" + *pigmentum*, "pigment"), due to its deep-yellow color, was thought to contain gold. The Greek for orpiment was *arsenikon*, from the Greek word *arsenikos*, which meant "male." (Metals, at the time, were believed to be of different sexes.)[11]

Closely related to orpiment, realgar has a brilliant red color. Well-formed realgar crystals can look so much like red gemstones that the mineral was often called "ruby sulfur" or "ruby arsenic." The name "realgar" came from the Arabic *rahj al-ġār* ("powder of the mine"), via Medieval Latin. The ancient Greeks called realgar σανδαράκη (*sandarákē*), which was first recorded as "sandarac" in English in the 1390s.

Alchemical Symbol for Realgar[12]

The yellow and red sulphides were well known to the ancients, and were described by Aristotle, Theophrastus[13] (Aristotle's pupil and successor, who lived around 300 BC and recognized both forms of arsenic, orpiment and realgar), Dioscorides (c 40-90 AD—a Greek physician, pharmacologist, and botanist), Pliny (23-79 CE—a Roman naturalist), and Strabo (c. 63 BC – c. 24 AD, a Greek geographer, philosopher, and historian), who mentioned a mine near Pompeiopolis,[14] where, because of its poisonous character, only slaves were employed.

The ancient Greek apothecaries also used realgar to make a medicine known as "bull's blood." The Greek physician Nicander described a death by "bull's blood," which matches the known effects of arsenic poisoning. Bull's blood was the poison that may have been used by Themistocles and Midas for suicide.[15] By the Middle Ages, it was well known that both orpiment and realgar could be combined with natron (sodium carbonate) to produce white arsenic (arsenic trioxide), a dangerously toxic substance. Realgar was not understood only as a poison, however. Along with orpiment, realgar was traded in the Roman Empire as a paint pigment.[16]

CHAPTER 9

Pseudo-Democritus
Besides being used as poisons or pigments, the arsenic ores were instrumental in many other endeavors of the early alchemists. One of the earliest recipes which mentioned realgar and orpiment was found in the works of Pseudo-Democritus, who traditionally has been purported to have been Bolus of Mendes, a philosopher and Neopythagorean writer of esoterica and medicine, in Ptolemaic Egypt around the 3rd or 2nd century BC. (This assertion has been seriously disputed, however. There is a consensus that some of the work in the Democritean corpus can be attributed to Bolos of Mendes, but other treatises are believed to be written by another person. Whoever the author was, his work has been placed by some scholars in the first century AD while others put forth an even earlier date, between 260 and 110 BC.)[17]

In his work *Peri asemou poiseseos (On the Making of Silver)*[18] Pseudo-Democritus described a recipe for tingeing copper with a silvery color using orpiment or realgar:

> *[Take] the mercury that comes from orpiment, or from realgar, or according to your knowledge, and make it solid as is customary; lay it on the copper [or] on the iron which have been purified with sulphur, and they will turn white. Whitened magnesia produces the same effect as well, and orpiment that has been turned inside out, and roasted cadmia, and unburnt realgar, and whitened pyrite, and white lead roasted together with sulphur. You will melt iron by adding magnesia, or half of sulphur, or a pinch of magnetite. For magnetite has affinity with iron. Nature delights in nature.*[19]

Pseudo-Democritus demonstrated that the effect of orpiment or realgar, properly prepared ("... according to your knowledge, and make it solid as is customary...") was known to the early alchemists. The orpiment or realgar could not simply be laid on the copper or iron. The arsenic sulfide would have had to have been converted to arsenious oxide (This process was often described as a whitening or fixing of arsenic, as will be seen in further discussion below), which was achieved by sublimation in air. In addition, the copper or iron was required to be "... purified with sulfur ...," which was, presumably, some sulfur-containing substance or solution. If copper was cleansed by boiling it with alum and acid solutions, or perhaps a sulfate solution, and was then melted with some arsenic compound, mixtures or alloys of copper and copper arsenide—which were white, lustrous, and silver-like in appearance[20]—would result. Alum was known to Pseudo-Democritus as evidenced by his discussion of the whitening of cinnabar (*"Make cinnabar white with oil, or vinegar, or honey, or bine, or alum . . ."*).[21] Copper and iron sulfides, also known by Pseudo-Democritus (*"Grind sori and copper flower together with unburnt sulfur; sori looks like azurite that easily peels off and is always found in misey. It is also called green copper flower. Roast it at moderate heat for three days, until it becomes a yellow drug."*),[22] could release sulfur dioxide on heating, and then react with water and air to produce sulfuric acid. The acid, then, would attack the non-metallic impurities thus purifying the copper.

Pliny the Elder (23-79)
The Roman naturalist Pliny was also well acquainted with the arsenic sulfides, especially sandarach, known as realgar. In Book XXXIV, chapters 55 and 56 of his *Natura Historia,* he described sandarach (realgar) and its medicinal value:

> *CHAP. 55.—Sandarach; Eleven Remedies.*
> *We have already mentioned nearly all the properties of sandarach. It is found both in goldmines and in silvermines. The redder it is, the more pure and friable, and the more powerful its odor, the better it is in quality. It is detergent, astringent, heating, and corrosive, but is most remarkable for its septic properties. Applied topically with vinegar, it is curative of alopecia. It is also employed as an ingredient in ophthalmic preparations. Used with honey, it cleanses the face and makes the voice more clear and harmonious. Taken with the food, in combination with turpentine, it is a pleasant cure for cough and asthma. In the form of a fumigation also, with cedar, it has a remedial effect upon those complaints.*

In chapter 56, Pliny used the term *arrhenicum*. Though this was a foundational word for the English arsenic (see the discussion above), it was most likely a sulfide of arsenic. The descriptive phrases "...the color of the finest gold..." and "...resembling sandarach, being less esteemed..." suggested that the sulfide referenced was most likely orpiment. In this same chapter, he mentioned a third kind of *arrhenicum* which has the color of gold and of sandarach. It is possible that this was also realgar since the color for realgar ranges from red to yellow-orange. Another possibility is that it was a mixture of orpiment and realgar since both occur together and realgar naturally decomposes to orpiment. Another clue to the identity of this third kind of *arrhenicum* was in the description that *arrhenicum* and the third kind were "scaly," meaning that they naturally divided into laminae. Orpiment's crystalline structure consists of sheets stacked in layers resulting in the scaly effect. The other (sandarach referenced in chapter 55) divided into fine fibers: "The last two kinds are both of them scaly, but the other is dry and pure, and divides into delicate long veins...." His description would be consistent with realgar which has a striated crystalline structure:

> *CHAP. 56.—Arrhenicum.*
> *Arrhenicum, too, is procured from the same sources. The best in quality is of the color of the finest gold; that which is of a paler hue, or resembling sandarach, being less esteemed. There is a third kind also, the color of which is a mixture of that of gold and of sandarach. The last two kinds are both scaly, but the other is dry and pure, and divides into delicate long veins. This substance has the same virtues as the one last mentioned, but is more active in its effects. Hence it is that it enters into the composition of cauteries and depilatory preparations. It is also used for the removal of hangnails, polypi of the nostrils, condylomatous tumours, and other kinds of excrescences. For the purpose of increasing its energies, it is heated in a new earthen vessel, until it changes its color.*[23]

Two early appearances of arsenic in the works of the alchemists of the Common Era came from the works of Agathodaemon (c. 300), a member of Maria the Jewess's school and an alchemist referenced by Zosimos[24] and Zosimos of Panopolis (end of the 3rd and beginning of the 4th century AD).

Agathodaemon (c. 300)

Agathodaemon (Greek: Ἀγαθοδαίμων, *Agathodaímōn*) was an alchemist in late Roman Egypt or Seleucid Syria,[25] known only from fragments quoted in medieval alchemical treatises such as the *Anepigraphos*,[26] which referenced Agathodaemon's works believed to be from the 3rd century. He was known for his descriptions of various elements and minerals, most notably his description of a method of producing silver and of a substance he created, which he called a 'fiery poison' (probably arsenic trioxide). He described the 'fiery poison' as being formed when a certain mineral (most likely realgar or orpiment) was fused with natron (naturally occurring sodium carbonate), and when dissolved in water gave a clear solution (arsenic trioxide). He pointed out that when a fragment of copper was placed into the solution, the copper turned a deep green hue, which further validated the suggestion that orpiment or realgar was used—as they are both arsenic ores, and the color achieved from the copper placed in the solution would have been consistent with the green color of copper arsenite ($CuHAsO_3$).[27]

In his writings, Agathodaemon purported that his teacher and only authority was Hermes[28] and that his work was based on the sayings of Hermes, particularly "The Stone is a Stone and not a Stone" because it was a stone in appearance but not in its property of dissolving. Further, "the Stone" was made from a single material, which dissolved and became clear water and pure spirit. This was the essential nature of stones according to Agathodaemon's reiteration of Hermes's teaching:

> *...the student should confine himself to the one secret-illuminating sentence. This 'Stone' by which the Work is performed is a Stone and not a Stone. This 'Noble Stone' which God has bestowed on us is not one of the ordinary stones, seeing that it melts and comes out as the Essential Nature (kiyan) of the stones—a Clear Water and a Pure Spirit. After being mixed with whatever is necessary and heated, it coagulates into the 'Etesian Stone,' through which alone Tincturing is possible. Copper, when treated as science prescribes, becomes Silver...*[29]

According to Agathodaemon, if the material was mixed with a substance, which he left unnamed, pulverized by fire, and evaporated to paste, it became a stone similar to copper burned in its own sulfur. If it was manipulated further, the "Stone" became leaf-like and many colored, and on further treatment with small quantities of liquid it became gold. Agathodeamon advised that much attention needed to be given to the extent of heat applied to purify the Stone, and described the operation as a sequence of colors obtained: red, yellow, white, black, and green. The final tincture was purple, of a sweet taste and fragrant odor. He also identified agents used in the process as a "gum" (kola) and "a fiery poison extracted by fire from the natures," most likely from distillation. As discussed above, the "fiery poison" was probably arsenic trioxide.

According to Dr. Robert P. Multhauf[30] (American science historian, b. 1919- d. 2004), the sequence of operations could be explained if it were assumed

Agathodaemon's starting point was to transmute copper. If the single material was realgar (arsenic disulfide) and if it were fused with natron or mercury, arsenious oxide, "a fiery poison," would result. Arsenious oxide was consistent with Hermes's teaching "a Stone which is not a Stone," because this oxide of arsenic was capable of forming both "a clear water" (solution) and "a pure spirit" (white sublimate). If it were mixed with vegetable oil (Agathodaemon's gum?) and heated, another sublimate formed (elemental arsenic) which, when applied to copper in the *kerotakis*,[31] imparted the copper with a silvery color.

Multhauf maintained that if, as Agathodaemon advocated, the applied heat were carefully controlled, the red realgar, fused with sulfur, could be converted into the yellow sulfide, orpiment. Other starting materials could have been copper and orpiment, if red referred to copper and yellow referred to orpiment. Either way, if the orpiment were fused with mercury or natron, the white sublimated oxide would result, which could be interpreted as "the Stone." Further, if the oxide were fused with oil or gum, a black sublimate—elemental arsenic— would result. Elemental arsenic, then, could be used as a "tincture" capable of imparting a silvery color to copper, an action of which the compounds of arsenic were incapable.[32]

Zosimos (end of the 3rd and beginning of the 4th century AD)

Another early supposed isolation of arsenic was that described by Zosimos of Panopolis (Greek: Ζώσιμος ὁ Πανοπολίτης; also known by the Latin name *Zosimus Alchemista*, i.e. "Zosimus the Alchemist") who was a Greco-Egyptian alchemist and Gnostic mystic. He was born in Panopolis (present-day Akhmim, in the south of Roman Egypt) and flourished ca. 300. He wrote the oldest known books on alchemy, which he called *Cheirokmeta*, the Greek word for "things made by hand."[33]

In his works, Zosimos supposedly described the isolation of arsenic and the converting of copper into silver (this was most likely copper arsenide, which would have been a white metallic-looking compound similar in color to silver). E. J. Holmyard (an English science teacher at Clifton College,[34] Bristol, England, b. 1891 – d. 1959), an historian of science and technology, paraphrased Zosimos's recipe:

> *The second 'mercury,' arsenic . . . can be obtained from sandarach [arsenic sulphide], by first roasting it to get rid of the sulphur, when the 'Cloud of Arsenic' [arsenious oxide] will be left. If this is heated with*

> *various [reducing] substances, it yields the second mercury [metallic arsenic], known as the 'Bird,'* [35] *which can be used to convert copper into silver [copper arsenide].* [36]

This recipe introduced additional terms and code words associated with arsenic. First, the recipe understood arsenic as a "mercury" and the metallic arsenic as a "second mercury." Also, the oxide of arsenic was referred to as the "Cloud of Arsenic." Holmyard identified Zosimos's sandarach as arsenic sulfide, which could be either As (II) sulfide, realgar, or As (III) sulfide, orpiment. If the sandarach was realgar, by which sandarach was also known, it would still be conceivable that the roasting of As_4S_4 would produce the same results. Also, since As_4S_4 easily decays to orpiment, it would also be possible that the actual substance would have been orpiment. (The dissolution of realgar would be a redox reaction that involved the oxidation of As (II) to As (III) and the direct release of sulfide.)[37] In either case, the roasting (oxidation) of orpiment or realgar would drive off the sulfur and produce arsenious oxide or arsenic trioxide according to the reaction equations: $2\ As_2S_3 + 9\ O_2 \rightarrow 2\ As_2O_3 + 6\ SO_2$ or $As_4S_4 + 7O_2 \rightarrow As_4O_6 + 4SO_2$. The sublimation of arsenious oxide in the presence of realgar or orpiment, then, would result in a sublimate of metallic arsenic ("the Bird") and the evolution of sulfur dioxide.[38] The metallic arsenic could then be used to convert copper to form copper arsenide ($AsCu_3$).[39] (It is also conceivable that arsenic oxide could be reduced to metallic arsenic using the proper reducing agent, such as charcoal, and described by $2As_2O_3 + 3C \rightarrow 3CO_2 + 4As$.)

From Pseudo-Democritus through Zosimos, much attention was given to naming and categorizing the arsenical substances—including the ores and the results derived from the alchemists' recipes. (Interestingly, the first isolation of elemental arsenic may have occurred during this time—though this is highly speculative. For instance, as John Emsley (an academic specializing in chemistry and Science Writer in Residence at Cambridge University, 1997-2002) noted below, in the discussion under Albertus Magnus, Albertus was believed to be the first to isolate arsenic. Holmyard, however, as pointed out above, believed Zosimos was the earliest to isolate arsenic. George Sarton attributed the discovery to Jabir ibn Hayyan—an early 9th-century Arab alchemist. The issue is that no definitive documentation exists that would clearly indicate that the discovery was actually made. What we have are conjectures based on the alchemists' recipes. The first published account of the discovery and the recognition of arsenic as an element came from Johann Schroder in 1649.[40]

The Book of the Treasure of Alexandria (9th century CE)

In the 9th century, a work of unknown authorship, *The Book of the Treasure of Alexandria* (BTA), offered several recipes involving arsenic. The treatise was probably a Harranite[41] compilation of earlier texts and formularies, which had been cited in one of the books belonging to the Jabirian Corpus,[42] namely, the *Book of the LXX*, which criticizes several forgeries — the BTA among them — for implying that alchemical secrets were hidden in inscriptions on temples, pyramids, royal treasures, and so on.

The book offered various recipes involving arsenic in what the work referenced as the preparation and manipulation of the "three elixirs." Several recipes involved the extraction of substances called "sharp waters" by untranslatable names such as *Ra'rasius* and *Triras*, that, according to the BTA, were important to the elaboration of the "three elixirs," which the BTA claimed were the "great stone" or the Philosopher's Stone. These "waters" were also instrumental in the purification and sublimation of arsenic, which in turn was necessary for the purification or whitening of copper:

> *Chapter on the purification of arsenic*
> *Melt arsenic in an iron pot on a fire and add the same amount of ground glass. If glass is melted first and added to arsenic only afterwards, know that it will be more efficient. Next, grind it with human urine for a full day until night and throw in seven times its measure of trirās. Leave it for one day and one night and remove it. Next, throw in enough vinegar to cover it, 1/3 of water of leek, plus 1/3 of vinegar and 1/3 of sweet water. Cook everything on a low fire until it becomes fairly dry. Wash the arsenic with vinegar and then with water and, when dry, set it aside in a vessel.*[43]

Since the recipe was concerned with purifying arsenic, the reference to arsenic was presumably to one of the sulfides of arsenic. The composition of the glass was unclear and its role, therefore, cannot be ascertained. The primary ingredient in urine, besides water, is urea ($CO(NH_2)_2$). The vinegar, of course, was acetic acid (CH_3COOH), but without knowing the compositions of *Triras*, water of leek, and sweet water, it is not possible to determine how this recipe worked. The BTA, however, offered another recipe for the purification of arsenic, which involved more inorganic substances:

Another chapter «on the purification of arsenic» easier than the first
Break yellow arsenic into small pieces the size of chick peas, put them in an iron pot on the fire, and melt. Feed it with ground kundur, that is to say, olibanum up to the weight of the former; next, feed it again with the same measure of ground olibanum and glass ; mix them; next, feed it with 1/6 of cinnabar and 1/6 of sal-ammoniac. Whoever manipulates this, in order to avoid the odor of cinnabar, must put a cotton soaked in tar in his nostrils. Upon removing [this mixture] from the fire, throw on it the equivalent of 1/6 of alum and 1/4 its weight of mercury. Then, grind it in a salaya with vinegar alone for a full day. Next, cover it with 10 times its measure of cow's milk and put it to rot in a well for 21 days. Upon removing it, grind it again with vinegar; next, wash it with vinegar and then with water seven times, until water comes out clean and bright. Put it to dry in the shade and set it aside.[44]

The yellow arsenic would have been orpiment. Olibanum[45] (also known as frankincense, an aromatic resin obtained from trees) has a complex composition: including acid resin (6%), soluble in alcohol and having the formula $C_{20}H_{32}O_4$, gum (similar to gum arabic) 30–36%, and incensole acetate, $C_{21}H_{34}O_3$.[46] The other ingredients included: cinnabar (HgS), sal-ammoniac (ammonium chloride, NH_4Cl), alum (most likely potassium aluminum sulfate), vinegar (acetic acid), mercury, and cow's milk (It is unclear what role this would play). In addition to instructions for the purification of arsenic, the recipe also warned the reader of the noxious odor and toxicity (the mercury component would have been extremely toxic) of roasting cinnabar by advising experimenters to plug their noses with cotton soaked in tar to avoid breathing any fumes.

The sublimation of arsenic (presumably a sulfide of arsenic) was described as follows:

Chapter on the sublimation of purified arsenic
You must prepare it with the equivalent of twice its [amount] of mercury, which gives the best results, a quicker sublimation, and more efficiency when used in the work you intend to employ it. If the sublimate is white at the beginning, it is what is desirable no matter whether the sublimate is not very white at the first time. Then take [the part] that has sublimated and grind it in a salaya with vinegar and 1/6 its

> weight of sal-ammoniac. Next, wash it with vinegar, which must be immediately taken away, and wash it with water several times until the water becomes clear and sweet. After that, grind [the remaining arsenic] with copper filings, the amount of which must be equal to half the weight [of the arsenic], and ground glass equal to 1/6 that weight. When everything is well mixed, add the equivalent of half the mercury and make it sublimate in the aludel. It will sublimate better than it did the first time with regard to its color and brightness, as long as neither problems nor wrong procedures happen during its preparation. If you are not content with its whiteness, repeat this last-mentioned task [until] it sublimates white, of a limpid whiteness. And be careful that the fire is not too high, as any excess might burn it, because the nature of arsenic is very close to the nature of sulfur. Thus, know this.[47]

Sublimation of the arsenic sulfide would produce the white arsenic oxide (As_2O_3), which the above recipe noted as a white sublimate. The treatise also warned not to have the fire too hot since the arsenic oxide would burn off similar to sulfur ("... the nature of arsenic is very close to the nature of sulfur ...").

Finally, several recipes were provided for the whitening of copper. Some of these recipes used only the generic term arsenic, not distinguishing yellow arsenic, orpiment, from red arsenic, realgar. (This may have been indicative that it did not matter which arsenic sulfide, orpiment or realgar, was produced or that the translator simply translated arsenic for both terms. The example of a recipe for whitening copper below, however, specified yellow arsenic which would imply that the arsenic sulfide involved was orpiment.)

> Chapter on another way [of whitening copper]
> Take 1/2 Ratl[48] of [mercury and put it in a salaya]; throw on top of it sal-ammoniac equal to 1/6 of its weight, ['isfidaj of lead] equal to 1/4 of the weight of the mercury and grind them well in the salaya. Sprinkle as much of the water called triras as needed and grind this for one day and one night. Then, throw [on the mixture] 1/4 of its total weight of dust of yellow arsenic and grind it for one day and one night, while sprinkling as much triras as needed for it to become pulverized and blended. Leave it in a well-made ceramic pitcher and seal it with clay of wisdom and [leave it] to dry. Next, seal it again

> *with clay of wisdom and dry it again. Seal it for the third time and dry it in the shade. Put it on a low fire for 7 hours. Upon breaking the vessel, its content will be cobbled and dry, like cinnabar, though its color is not red like cinnabar's. Then, grind this with hen's egg white for a whole day, dry it in the shade and [next] sprinkle on it strong vinegar and grind it with vinegar for one more day. Wash it seven times with vinegar and seven more times with sweet water, dry it and add the equivalent of its weight of mercury stirring it while you pulverize. Then, sublimate it in the aludel. Thence comes out a stone the color of naqra[49], but less white than it.[50]*

"Clay of wisdom" (also known as "philosophers' clay") may have consisted of two-thirds clay free of stones and one-third a mixture of dried dung and chopped hair,[51] that was used as a luting[52] material. (Given the untranslatable term, "*Triras*," and the possibility that "hen's egg white" could either be actual egg white or code for another substance, the recipe does not lend itself to a full explication of the role of orpiment in the procedure.)

Al-Razi (Abu Bakr Muhammad ibn Zakariya al-Razi c. 865-923)

Al-Razi, a Persian physician, philosopher, and alchemist, in his *The Book of Secrets,* "Section One: What One Must Know About Substances," lists the substances required for the "chemical art." Al-Razi's taxonomy defined substances as animal, vegetable, and mineral (a scheme which is used colloquially to this day). The classification of mineral was further divided into six groups: spirits, metals, stones, vitriols, boraxes, and salts. Arsenic was considered both a spirit and a stone.

> *Concerning substances, there are three classes: animal, vegetable, and mineral. But the minerals fall into six groups: spirits, metals, stones, vitriols, boraxes, and salts...*
>
> *Of spirits, there are four: mercury, sal ammoniac, sulfur, and arsenic sulfide.*
>
> *Of metals, there are seven: gold, silver, iron, copper, tin, black lead, and Chinese iron*[53]

CHAPTER 9

> *"Of stones, there are thirteen: marcasite, magnesia, iron ore, tutia, lapis lazuli, malachite, turquoise, hematite, white arsenic, kohl, talc, gypsum, and glass."*[54]

He further differentiated arsenic sulfide (the Arabic was *"Zarnich"* which has been translated as arsenic sulfide) into various kinds in "Part Two: Distinguishing the Good and Bad Varieties" under "The Types of Spirits":

> *There are various kinds. Among them is a greenish kind, mixed with earthy rocks; this is the most worthless of the arsenics. Then one that is yellow and impure, mixed with earth; this is used in baths. Then a saturated yellow, flaky, gold-colored one; this suits us and is excellent. Then a yellow one, mixed with red; this is likewise excellent for our work. Another, that has grey [sic] flecks, is not useful to us. A red one, pure red and flakey, is especially excellent for our work.*[55]

The kind described as yellow mixed with red may have been a mixture of orpiment and realgar since these often occurred together, or it could have been just realgar since its coloring varies from orange-red to red. The saturated yellow, flakey, and gold-colored was most likely orpiment (Orpiment is yellow to golden-yellow and has been described as flaky or scaly). The last one described as pure red and flakey was most likely realgar.

The "greenish kind" may have been a copper or copper-iron arsenate ore. The most common arsenic mineral sources in Iran (Persia) were arsenopyrite (FeAsS) and arsenian pyrite, and travertines (a form of limestone) contained high concentrations of As.[56] Copper was mined in ancient and pre-modern Iran, along with iron, gold, lead, zinc, and silver,[57] and the copper arsenates would form in the copper deposits from the oxidation of copper ores and arsenopyrite. The most common secondary copper arsenate that would have formed in the oxidized zone of copper deposits would have been olivenite.[58] (Could the greenish color be that of olivenite, Cu_2AsO_4OH, which has an olive green coloration, or of lammerite, $Cu_3(AsO_4)_2$, which is dark green in color? Another, less common, copper-iron arsenate mineral, known as Arthurite, also has a greenish color.) A more common copper mineral containing arsenic, however, would have been enargite, Cu_3AsS_4, with a steel-gray to black coloration, which would not match al-Razi's "greenish" qualification. But enargite has a common oxidized alteration—scorodite, $FeAsO_4 \cdot 2H_2O$ (pyrite was most likely the source of the iron), which can be green in color. In addition to being a valuable source of copper, enargite was also mined for several precious metals including gold and silver.[59]

The kind of arsenic sulfide with the gray flecks may have referred to arsenopyrite which is steel-gray to silver white in appearance. (The identification of possible sources with al-Razi's color scheme for the arsenic classified as spirit is conjecture by the author.)

Al-Razi distinguished arsenic sulfide, a spirit, from white arsenic, a stone ("Part Two: Distinguishing the Good and Bad Varieties" under "The Types of Stones"), which was further defined as being of two kinds—yellow and white: *"White arsenic. There are two kinds, one yellow and one white. It is brought out of silver mines...and is the smoke of silver."*

The white arsenic was derived from the processing of arsenic-laden silver ore.[60] The references to the silver mines and "the smoke of silver" indicated an extractive process whereby the mined ore was refined through a smelting operation. (The exact nature of the yellow and white kinds of the white arsenic, attributed to silver extraction, is unclear. If, however, enargite was mined for gold and silver, then a possible explanation emerges. Above 550°C enargite begins to decompose, and the arsenic is removed as As_4S_4 which could further decompose to As_2O_3 and/or react with oxygen to yield $2\ As_2S_3 + 9\ O_2 \rightarrow 2\ As_2O_3 + 6\ SO_2$ or $As_4S_4 + 7O_2 \rightarrow As_4O_6 + 4SO_2$. In other words, both arsenious oxide, which is white, and orpiment, which is yellow, could be present from the processing of silver.)[61]

In addition to categorizing and describing arsenic, al-Razi expounded on how arsenic sulfide could be prepared for medicinal purposes and the whitening of copper. "The Chapter on the Procedures of Sulfur and Arsenic Sulfide" detailed the procedures for the sublimation and purification of arsenic and enumerated various procedures for the use of arsenic in coloring copper to various hues of silver—from whitish silver to grayish silver. Al-Razi methodically laid out this chapter, which listed the types of procedures—roasting, washing, boiling, sublimation, and "...making the essence visible..."—and the substances to be used in the recipes:

> *"The substances that sulfur and arsenic sulfide are treated with are copper acetate, quicklime, lime, iron filings, copper, tin, and black lead, vitriol, salt, white lead, lead oxide, glass, soda, talc, and sea foam, burnt copper, mashaquiniya,[62] white brick, the ash of gall nuts, of oak, and of carob. One of these is mixed together, and (the*

sulfur and arsenic sulfide) are treated with them. Also, waters can be added to them, and with them, simple and compound medicines are ground. These (waters) are vinegar, salt water, alum water, vitriol water, urine, sal ammoniac water, sour milk, and the acid pressed from lemons, soda water, lime water, and such." [63]

An example of a recipe for the sublimation of arsenic (orpiment) concluded that the arsenic would not have been properly prepared until a white smoke appeared, which would have been volatized arsenic trioxide:

"The secret of treating sulfur and arsenic sulfide is, however, that you put as much of their powder as you wish into a kettle and set the lid... coated with clay on top and seal the connection, after you have made a hole in the lid. Then go with it to a place where no one will notice the smell, in the desert or elsewhere ... and dig a pit for it in the earth and light a medium fire in it, set the kettle on the fire, and observe the smoke that comes out. You leave the kettle on the fire, as long as the smoke comes out black and yellow; however, when it begins to come out white, close the hole and take the kettle from the fire, so that it can get cold. Take the contents out, grind it fine, and sublimate it in a house or in your dwelling or where you will, because it will not harm you (now)." [64]

Moreover, al-Razi knew that arsenical minerals, if heated with iron, could produce a silvery iron–arsenic alloy. This alloy was not pure silver, but was like silver in appearance, which encouraged the belief that silver was somehow being formed, and led to numerous recipes for iron–arsenic alloys that promised the transmutation of iron into silver. For example, one such recipe came from the *Secret of Secrets*, where he described heating iron with *zarnich* (arsenic tri-sulfide, orpiment, As_2S_3):

Recipe 48. Another Method.
Take filings of iron, as much as you want, grind them with the same quantity of yellow zarnich and put them in a clay jug. Roast in the oven to just below melting point; afterwards, you make it into a powder with its sixth part of natron, and sprinkle it with as much oil as will unite with it; then it will melt like a metal, like Chinese iron.

The natron (sodium sesquicarbonate) would have acted as a flux, keeping the surface of the iron clean. (The comparison to "Chinese iron" is confusing, since this would have been cast iron, which melts at 1150°C,[65] while the melting point of a mixture of iron and iron arsenide (Fe_2As) is 840°C.)

Al-Khwārizmi (Muḥammad ibn Mūsā al-Khwārizmī, d. ca 997)

Al-Khwārizmi was a Persian polymath from Khwarazm,[66] who produced influential works in mathematics, astronomy, and geography. Between 976 and 980, al-Khwarizmi prepared the medieval Arabic encyclopedia entitled *MafaatiiH al-'uluum*, or *The Keys to the Sciences*, a reference work, which was dedicated to Abu al-Hasan 'Ubayd Allah ibn Ahmad al-'Utbi, vizier to the Samanid Nuh II ibn Mansur[67] (976-997). Intended as a manual for scribes and secretaries, the work provided the definitions and explanations of the technical terms which were coined and employed within the various disciplines at the time.

In regard to alchemy, al-Khwarizmi's works provided definitions, explanations, and categorizations, such as the alchemical distinction between "bodies" and "spirits," which he understood to be based on their reaction to the application of heat. The bodies included the seven metals: gold, silver, iron, copper, lead, tin, and quicksilver, whereas the spirits were sulphur, arsenic, mercury, and ammonia. He also listed different kinds of salts, ammonia, borax, vitriols, marcasite, magnesium, zinc, and a number of other naturally occurring substances which included gems (e.g., lapis lazuli, turquoise), kohl, ratsbane, colophony, arsenic, and the magnet. In "The Ninth Chapter of the Second Discourse: On Alchemy, section two—About This Craft," of his *The Keys to the Sciences*, al-Khwarizmi categorized arsenic:

> *The spirits: They are sulphur, arsenic, mercury, and ammonia. The bodies were so named because they remain stable when exposed to fire, but these are called 'spirits' because they 'fly' [evaporate] if touched by fire...arsenic of several types: red, white, and green; the green is the most deadly; the best of them is the 'flaky' (Safaa'iHii)...*[68]

In "Section Three—On the Operations and Preparations of These Things," al-Khwarizmi provided an additional reference to arsenic:

> *The mineral kinds of stone are gold, silver, lead, and tin. Among the spirits are mercury (al-ziibaq), arsenic, sulfur, and ammonia. Alkali*

> *of arsenic is the soul of whiteness, sulfur is the soul of redness, and mercury is the spirit of the two together. The elixir is composed of body and spirit.*[69]

In his taxonomy, al-Khwarizmi categorized arsenic as red, white, green, and "flaky" as opposed to the traditional red (realgar) and yellow (orpiment). The white arsenic most likely would have been arsenic oxide, the green could possibly have been a copper arsenate mineral (see the discussion on al-Razi above), and the "flaky" type could have been a reference to orpiment (see the above discussion on Pliny). Alkali of arsenic may have been another reference to arsenic oxide since the oxide is white and can be used to "whiten" (e.g., copper), which would be consistent with the reference "…the soul of whiteness.…"

Albertus Magnus (Albert the Great, 1193–1280)

Albertus Magnus was a German Dominican friar, philosopher, scientist, and bishop, who was a prolific writer. In *De Libellus de Alchimia*, he defined arsenic as one of the four spirits of metals, which could be used to color or tinge metals. In the tenth chapter, he wrote:

> *Note that the four spirits of metals are mercury, sulfur, auripigmentum or arsenicum, and sal ammoniac. These four spirits color metals white and red, that is, in gold and silver: yet not of themselves, unless they are first prepared by different medicines for this,*[69] *and are not volatile, and when placed in the fire, burn brilliantly. These spirits fashion silver from iron and tin, or gold from copper and lead.*[70]

Auripigmentum denoted a yellow mineral, impure arsenic (III) sulfide (As_2S_3), and *arsenicum*, unless specified otherwise, usually had the same meaning. Observed differences in color, however, designated As_2S_3 as *arsenicum citrinum*, which distinguished it from *arsenicum album*, As_2O_3, and *arsenicum rubrum*, As_4S_4.

The next paragraph was noteworthy since it demonstrated a vacillation between tingeing (a surface treatment) and transmutation (a change in a metal's identity).[71] After he discussed changing the color and the questionable quality of the gold, Albertus Magnus went on to state that the gold would last forever even after experiencing various operations:

Thus, as I shall say briefly, all metals may be transmuted into gold and silver, which are like all the natural metals, except that the iron of the alchemist is not attracted by adamantine stone and the gold of the alchemist does not stimulate the heart of man . . . But it is evident that in all other operations, as malleation, testing, and color, it will last forever. From these four spirits, the tincture is made, which in Arabic is called elixir and in Latin, fermentum.

Later, in the section of the work concerned with the fixation of powders, he described a method which may have isolated elemental arsenic from a compound by heating soap (oil of tartar—a saturated solution of potassium carbonate, K_2CO_3) with arsenic tri-sulfide:

A second way [to fix powders] is with the imbibition[72] of oil of tartar. However, you can do it this way: take sublimed arsenicum or sulphur or auripigmentum, and crush over the stone with oil of tartar, until all becomes liquid. Then place in a glass phial in ashes, which have been sifted through a fine sieve, and place the vessel with the ashes over a distillation furnace, and apply the fire very slowly as [is done] in masticating, lest the vessel be broken. After heating the glass, increase the fire; then dry the medicine in an open vessel, if you wish, but it is better [to do it] in a closed one. Place above it an alembic which collects the water distilled from it, because [this distillate] is useful for many things. When the medicine is dry, the vessel is to be broken, since you cannot empty it otherwise, and you will find the powders hardened like stone. This has to be well ground as before with the distilled oil [of tartar]. Using the same procedure, again break the glass, remove, grind well, place in another ampulla[73], and set [it] in a warm dung pit for seven days, and then it will be dissolved into a liquid. Then place the vessel in warm ashes and heat with a slow fire, then you will have the spirits fixed . . . And of this powder, add one part to fifty parts of calcined Iron or Copper . . .[74]

The "...spirits fixed..." may have indicated that he believed he had isolated metallic arsenic (It is unclear whether Albertus actually observed the free element). Nicolas Lemery (17 November 1645 – 19 June 1715), a French physician and chemist, did observe the formation of arsenic using a recipe that involved heating a mixture of arsenic oxide, soap, and potash. (The first authenticated report of the isolation of elemental arsenic was by Johann Schroeder, a German pharmacist, in 1649, who prepared arsenic by heating its oxide with charcoal.[75] The reaction equation would have been $2As_2O_3 + 3C \rightarrow 3CO_2 + 4As$.) In Albertus's recipe above, he began with a sublimed *arsenicum*—commonly referred to as the oxide of arsenic—and mixed it with oil of tartar. Lemery's reactants also started with arsenic oxide, and then were mixed with soap and potash (a concentrated solution of potassium carbonate, K_2CO_3). In the Middle Ages, soap was most likely made from a mixture of animal fat and an alkali presumably from wood ash such as caustic potash (potassium hydroxide, KOH, also known as salt of tartar). Oil of tartar, then, would have provided the same ingredients as Lemery's recipe. (Potassium hydrogen tartrate—oil of tartar, formula $KC_4H_5O_6$, a byproduct of the fermentation process in winemaking—was processed from the potassium acid salt of tartaric acid, a white, crystalline organic acid that occurs naturally in many fruits such as grapes.)[76] The heating of Albertus's mixture, then, could have produced the same result as Lemery's recipe—metallic arsenic. In addition, the recipe indicated that the vessel was sealed, except for small openings to relieve any pressure—"... *the vessel is to be broken, since you cannot empty it otherwise*. . . ." In this case, given a high enough temperature, the oil of tartar would burn, emitting carbon monoxide, which would react with the arsenic oxide vapor and result in the formation of arsenic. Another clue was found in the last sentence of Albertus's recipe. If he had isolated arsenic, then the end of the above recipe, which instructed to add the resultant powder to iron or copper, could have been an alloying of those two metals.

In line with the above argument for Albertus Magnus isolating arsenic, John Emsley (an academic specializing in chemistry and a Science Writer in Residence at Cambridge University, 1997-2002), also argued that Albertus may have produced the arsenic metal by heating vegetable oil mixed with arsenic trioxide or perhaps by substituting charcoal for the vegetable oil.[77] If Albertus did use charcoal, the reaction would have taken place in a sealed vessel, which would have presented an oxygen-starved environment. Under this condition, the charcoal would not have burned completely to carbon dioxide as it would have in open air.

On the other hand, if *arsenicum* would have referred to realgar (orpiment was already indicated by the term auripigmentum), or if the vessel was not thoroughly sealed, or if the reaction product was added to copper or iron only to tinge the two metals, then these points would counter the above argument for the isolation of arsenic. Further, nowhere does Albertus Magnus document that he knew he had isolated arsenic. (Of course, the possibility exists that he did produce it but did not recognize it.)

(Further discussion of claims for or against crediting Albertus Magnus with the isolation of elemental arsenic is beyond the scope of this essay; it is left to the reader to pursue the disputations and challenges. For instance, George Sarton [31 August 1884 – 22 March 1956, a Belgian-American chemist and historian considered the founder of the discipline of the history of science] postulated that Jabir ibn Hayyan isolated arsenic before 815.[78] Recent scholarship on the Jabir Corpus, however, questions the authorship and dating of these writings, which would present challenges to Sarton's position. Another early disputant was none other than Georgius Agricola—see the discussion of Georgius Agricola below.)

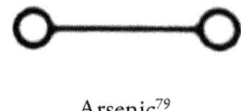

Arsenic[79]

Al-Jawbari (Abd al-Rahman ibn 'Omar al-Din al-Dimashki, early 13th century)

Arsenic was not only used to "whiten" copper or to make pigments, it was also used to deceive. In the early 13th century, an Arab author known as al-Jawbari (This was his nickname because he was born in Jawbar, near Damascus.) wrote *Kitab al-mukhtar fi nashf al-asrar al Dimashki*, which exposed the fraudulent practices of wandering fake alchemists and money changers. One of the deceptions of these charlatans involved arsenic. The false alchemists would melt metals with arsenic, converting the mixture into what appeared to be an ash. Then the tricksters would claim to be able to make an "elixir," which would be poured on the ash in a crucible. The mixture would be heated, causing the metals to be separated from the arsenic in the ash and, thus, to reappear.[80]

Pseudo-Geber (13th century)

Pseudo-Geber (Paul of Taranto) wrote extensively on the preparation of arsenic for use in other alchemical operations in the chapter "Of the Alchymie of Arsenick" in his *Summa Perfectionis* (R.A.M.S. translation). Like the tradition before him, he also identified the more common forms of arsenic ores by the color scheme of yellow and red with his terms "Citrine" and "Red" for orpiment and realgar, respectively. In agreement with Pliny, he further qualified that the best arsenic ore was scaly (divided into laminae), and easily split (*scissile*). Orpiment and realgar both met those requirements—opriment's structure was laminated as sheets stacked in layers, and realgar's structure was like that of fine fibers. The qualifier "Lucid" may have referred to orpiment's and realgar's optical characteristic of transparency, which allows light to pass through without scattering:

> *Of Arsenic, there is a Citrine and a Red, which are profitable in this art, but the other kinds not so: Arsenic is fixed as Sulphur, but the sublimation of either is best from the Calx of Metals...The best kinds of Arsenic are the Scissile (cut or split easily), the Lucid, and Scaly.* [81]

To prepare orpiment or realgar for later use, Pseudo-Geber described a method which presumably produced arsenious oxide. According to the recipe below, after boiling powdered arsenic sulfide (either orpiment or realgar) in vinegar (CH_3COOH), it was added to a mixture of calcined copper, alum (potassium aluminum sulfate), salt (NaCl), and vinegar, which was boiled and then sublimed. From the description—white, dense, clear, and lucid—arsenic trioxide was the result, which could be used later for "the Work":

> *Of the Preparation of Arsnick. After its Compeer Arsenick is beaten to Powder, it must be boiled in Vinegar, and all its combustible Fatness extracted, and it then dried. Then ℞ of Copper, calcined, lib. 1. Of Allom calcined ½ a pound, and of Common-Salt prepared as much as of the Allom. Mix these with your Arsenick prepared, and having ground all well together, moisten the Mixture with distilled Vinegar (that it may be liquid) and boil the same, as you did in Sulphur; and then sublime it in an Alludel (without an Alembick) of the height of one Foot. Gather what ascends white, dense, clear, and lucid, and keep it; because it is sufficiently prepared for the Work.* [82]

Guillaume Sedacer (d. 1382)

The use of arsenic as an alloying agent in iron was known not only to the Arabic alchemists (see the discussion on al-Razi above). In the fourteenth century,

Guillaume Sedacer, a Catalan Carmelite, who had a great interest in alchemy, medicine, and astronomy,[83] transcribed a recipe for "fusion of iron" involving arsenic:

> *In Damascus, it is cast like this: take very thin sheets of iron and heat them and quench them 40 times in a wash made of the ashes of nettles until they are as white as silver. Then coat them with a powder of fresh pomegranates and myrobalan[84], close it up [like a] prison, and dissolve it on a great fire; do this three times. Then it may be cast at once, fused with the fixed sublimate of arsenic called "white sulphur of the philosophers" [presumably arsenic trioxide, As_2O_3], and to one part of the aforesaid arsenic, add 200 parts of the aforesaid fused iron; therefore, this will be transmuted into a silver better than that [which is] mined.*[85]

Paracelsus (c. 1493 – 24 September 1541)

Paracelsus, in *De Naturalibus Rebus*, differentiated arsenic as that which occurred naturally and that which was made by the Art (a reference to the alchemical art) through transmutation. He further defined the naturally occurring arsenics into native arsenics—arsenic minerals—and arsenics derived from metals (metal ores may have been meant here). From there, he described the kinds of arsenic by color and application: white arsenic for medicinal purposes, and yellow and red for the alchemists' investigations of the transmutation of metals:

> *Concerning the kinds of arsenic, it is to be noted that there are those which flow forth from their proper mineral or metal and are called native arsenics. Next, there are arsenics out of metals after their kind. Then, there are those made by Art through transmutation. White or crystalline arsenic is the best for medicine. Yellow and red arsenic are utilized by chemists for investigating the transmutation of metals, in which arsenic has a special efficacy.*[86]

Paracelsus did more than classify arsenic, however. He also explained the nature of arsenic as having three "spirits"—one that was volatile and whitened copper, another that was crystalline with a sweet taste, and a third that was used for tingeing:

CHAPTER 9

> *The Aurora of the Philosophers, Chapter XI—Concerning the True and Perfect Special Arcanum of Arsenic for the White Tincture. Arsenic contains within itself three natural spirits. The first is volatile, combustible, corrosive, and penetrating all metals. This spirit whitens Venus (copper) and after some days renders it spongy. But this artifice relates only to those who practice the caustic art. The second spirit is crystalline and sweet. The third is a tingeing spirit separated from the others mentioned before. True philosophers seek for these three natural properties in arsenic with a view to the perfect projection of the wise men. But those barbers who practice surgery seek after that sweet and crystalline nature separated from the tingeing spirit for use in the cure of wounds, buboes, carbuncles, anthrax, and other similar ulcers which are not curable save gentle means.*[87]

His classification of arsenic and his discussion of arsenic's nature singled out the white, crystalline, sweet-tasting kind of arsenic for medical treatment. In particular, this arsenic, which was separated from the "tingeing spirit," was used for the cure of wounds, buboes, carbuncles, anthrax, and other similar ulcers. Knowing that arsenic was toxic, Paracelsus nevertheless defended the medicinal use of the substance. By means of laboratory procedures, he affirmed that something beneficial could be made out of something potentially harmful in itself, and he specifically offered arsenic as an example:

> *As it is possible to make something bad from something good, it is also possible to make something good from something bad. [Thus] no one should chastise a thing who does not recognize its transmutation and who does not know what separation does. (Paracelsus, Septem Defensiones)*
>
> *For example, the use of arsenic, one of the chief poisons. A drachma can kill a horse. But "burn it with sal nitri [potassium nitrate] and it is no longer poison." The difference is in the preparation, in the action of completing and transmuting nature.*[88]

In order to make use of arsenic, Paracelsus provided several recipes for fixing[89] it (presumably arsenic oxide). One such recipe was from *De Naturalibus Rebus*:

One recipe for the fixation of arsenic:—Take equal parts of arsenic and nitre. Place these in a tigillum[90], set upon coals, so that they may begin to boil and evaporate. Continue till ebullition[91] and evaporation cease, and the substances shall have settled to the bottom of the vessel like fat melting in a frying-pan; then, for the space of an hour and a half (the longer the better), set it apart to settle. Subsequently pour the compound upon marble, and it will acquire a gold color. In a damp place it will assume the consistency of a fatty fluid.[92]

In this recipe arsenic, presumably arsenic sulfide, was heated with potassium nitrate (nitre, KNO_3) in a crucible. Paracelsus did not specify if, by arsenic, he meant the sulfide or the oxide, and he did not indicate whether the potassium nitrate was in solution. (Without further information, it is unclear what was actually produced by this recipe.)

Like the Arabic author al-Jawbari, Paracelsus also warned of the deception perpetrated by others who claimed to have transmuted copper into silver. He recognized that arsenic could be used to tinge or change the color of copper to a silvery color:

Some have taken arsenic several times sublimated, and frequently dissolved with oil of tartar and coagulated. This they have pretended to fix, and by it to turn copper into silver. This, however, is merely a sophisticated whitening, for arsenic cannot be fixed unless the operator be an Artist, and knows well its tingeing spirit. Truly in this respect all the philosophers have slept, vainly attempting to accomplish anything thereby.[93]

Further on in the chapter, "Concerning the True and Perfect Special Arcanum of Arsenic for the White Tincture," he was more specific in his criticism of those alchemists—including Geber, Albertus Magnus, Aristotle the Chemist, Rhasis, and Polydorus—[94] in regard to the proper preparation of the white tincture:

Some persons have written that arsenic is compounded of Mercury and Sulphur, others of earth and water; but most writers say that it is of the nature of Sulphur. But, however, that may be, its nature is such that it transmutes red copper into white. It may also be brought to such a perfect state of preparation as to be able to tinge. But this is not done in the way pointed out by evil sophists as Geber in "The Sum of Perfection," Albertus Magnus, Aristotle the chemist in "The Book of Perfect Magistry," Rhasis, and Polydorus; for those writers, however many they be, are either themselves in error or else they write falsely out of sheer envy and put forth receipts whilst not ignorant of the truth.[95]

Further, in his attack, Paracelsus revealed how other alchemists had viewed the nature of arsenic. Some saw in arsenic the qualities of Mercury (the volatile and metallic nature of arsenic) and Sulfur (the combustible nature of arsenic). Arsenic, understood as a compound with sulfur (the element as opposed to the principle—Sulfur), would easily be seen as having these qualities. Paracelsus dismissed these views, however, as not as significant as the ability of arsenic to whiten copper, and that "...the way pointed out by [these] evil sophists..." would not accomplish that objective. (It must be remembered Paracelsus disdained the traditions that preceded him and believed he had the true understanding of alchemical and medical subjects. Hence, his epitaph "Bombastus.")

Though Paracelsus, like earlier alchemists, addressed such applications of arsenic as tingeing, whitening copper, and transmutation, a significant focus of his efforts was on the medicinal qualities of arsenic. One of his most important contributions was his demonstration of how even a toxic substance such as arsenic, if properly employed, could be beneficial in medical applications.

From antiquity to the early 16th centuries, the development of the chemical knowledge of arsenic was rather convoluted and filled with arcane language, swirling eddies of recipes to achieve various ends, a sometimes obtuse language, and arcane theories—a mixture without a unifying structure. (This is understandable given the alchemists had to often invent terms and descriptions, and were often unaware of others' work.) Beginning in the 16th century, however, there were those who, through their work, provided clearer, more consistent, and practical descriptions of arsenic and its compounds. One such person was the metallurgist Georgius Agricola.

Georgius Agricola (24 March 1494 – 21 November 1555)

Agricola, a German mineralogist and metallurgist, advanced the classification of the arsenic ores by specifying more information about their physical characteristics, their locations along with their geological descriptions. He first identified two genera of arsenic: orpiment (*auripigmentum*) and realgar (*sandaraca*), which he attested were closely related and almost always appeared together in the same ore veins. He added to their description by explaining how they responded to heat—both appeared as if burnt. But he noted that orpiment and realgar burned with greater difficulty when compared to other substances containing sulfur or bitumen.[96]

He continued by distinguishing orpiment from realgar, affirming that there were three qualities of orpiment. The first had the nature or characteristic of an unguent or ointment. It was more oily or greasy than realgar. In addition, orpiment had an acid taste and, when burned, gave off a sulfur odor which was less intense than that given off by realgar. Citing Dioscorides and Vitruvius,[97] Agricola noted that orpiment and realgar were commonly found together, particularly in Hellespond, Mysia,[98] in Pontus[99] on the Hypanis river, in Asia Minor along the boundary between Magnesia and Ephesus, and in Cappadocia.[100]

In addition, Agricola specified that orpiment varied in color and form. One variety had the color of gold and was composed of thin layers resembling scales that appeared to be stacked on top of each other (see the discussion above regarding the structure of orpiment) and cleaved in a fashion similar to gypsum. This was orpiment of the finest quality. In regard to the second quality, Agricola confirmed Theophrastus, who described a variety of orpiment as having the color of gold and occurring as a powder. (This may have been a reference to a form of realgar which appears as a yellow powder.) He also described another variety of orpiment as having a color close to realgar and occurring in the form of lumps. Agricola further referenced Dioscorides regarding this variety by identifying Pontus and Cappadocia as locations where it could be found.

In addition to the natural varieties of orpiment, Agricola identified two artificial species. One was white with pale yellow or red veins coursing through it. The other variety was typically yellow but varied from black to pale red with yellow veins running through it. These were produced by adding salt into the cleavages of native orpiment.[101]

Agricola's second genera of arsenic, realgar, also varied in color from deep red—similar in color to cinnabar—to less red and, in parts, to pale yellow. Realgar was found in two different forms, as lumps—the most common form, or as powder. Agricola went beyond providing the physical characteristics in his classification of realgar. He also included the responsiveness of the arsenic sulfide to processing. The best quality, he attested, was that which did not require grinding or sieving but occurred naturally as a powder. Realgar had a sulfurous odor, but when it was crushed, the odor became stronger. When it was burned, the odor was even stronger, and noticeable yellow fumes were produced. Finally,

Agricola averred that the very best quality realgar, when pure, was a deep red in color, had an odor of sulfur without being ground or burned, could be crushed with ease, and was found in Pontus.

In regard to processing, Agricola shared important observations for treating ores by describing the color of the fumes given off when the ore in question was heated. These observations offered additional information for identifying orpiment and realgar. The color of the fumes emitted, when the ore was placed on a hot shovel or iron plate, indicated what flux was to be used for smelting. Moreover, the color indicated what substances were contained within the ore. For example, blue fumes signified that the ore contained azure; yellow fumes indicated orpiment; red, realgar; green, chrysocolla;[102] black, black bitumen;[103] white, tin; white with green patches, tin mixed with chrysocolla. If the middle part of the fume was yellow and other parts green, the ore, according to Agricola, contained sulphur.[104]

In addition to describing physical characteristics, location of occurrence, and responsiveness to mechanical working, Agricola wrote of artists and others who employed realgar in their work. He was especially interested in painters, potters, and physicians. Painters used realgar in the pigments of their paints. Potters used it to draw red lines around their pottery. Since the compound had the property of burning, physicians often used it in medical treatments. Agricola elaborated on this last application: If realgar was consumed with liquids or resins, it was thought to cure fevers, and, if mixed with honey, purulent expectorations would be cured. Further, if it was mixed with resin and burned, the inhaled fumes cured asthma and chronic cough. Hoarseness could be improved if the mixture of realgar and resin was licked or swallowed.[105]

It is odd that Agricola did not seem to be aware of metallic arsenic. If Albertus Magnus actually isolated arsenic, it would be reasonable to assume that Agricola—who was familiar with Albertus's works—would have mentioned it. But, according to Agricola, Albertus Magnus only referred to arsenious oxide, which resulted from sublimation. Agricola argued that the term *"arsenicum,"* as used in the early writings of the alchemists, typically referred to a "white variety," which meant the oxide sublimated from orpiment,[106] but he noted that realgar itself could also be sublimed from orpiment:

> *Realgar can be made from orpiment in the following manner. Medium-small particles of the latter are placed in an earthen jar and then the mouth of it is sealed. The jar is placed in a furnace for five hours and the mineral will then have the color realgar.*[107]

Agricola mentioned four minerals with an arsenic base: *Auripigmentum* (orpiment), *Sandaraca* (Realgar), *Arsenicum* (artificial arsenical oxide or white arsenic), and *Lapis subrutilus atque splendens* or Mispickel (Arsenopyrite, Fe-AsS). In addition, he also identified the cobalt-arsenic minerals of *Cobaltum cineraceum* (Smallite, $CoAs_2$) and *Cobaltum ferri colore* (Cobaltite, CoAsS). [108]

Martinus Rulandus (1532-1602)

Martinus Rulandus was a German physician and alchemist and a follower of Paracelsus. He summarized the terms used by Paracelsus and the Paracelsians to reference arsenic, along with many other terms, in a lexicon published posthumously in 1612. Accordingly, he listed "Artaveck and Artaneck" (arsenic) which consisted of three species—white, yellow, and citrine. The yellow was identified as orpiment which was further described as a golden dye and crystalline arsenic. The red was what the Greeks referred to as *sandaraca*, which was of two kinds, rough (presumably, meaning natural or native) and manufactured. The rough was described as red arsenic mixed with brimstone, and the latter as vermilion arsenic (possibly arsenic mixed with cinnabar).

Rulandus further designated realgar as red orespiment and defined it more properly as a "mineral smoke," (which he failed to explain further). In addition, he discussed the metaphorical use of the term by the Paracelsians. Metaphorically, they saw realgar as a poison of the body which was the cause of ulcers. In this very broad view, there were four kinds of realgar, each corresponding to one of the four elements: Realgar on the Surface of the Water, Arsenical Realgar of the Earth, Terebinthine,[109] Realgar of the Air, and Saturine Conjunction, the Realgar of Fire.[110]

By collecting the terms for various substances, operations, and apparatus, Rulandus provided a valuable tool for the history of the knowledge of arsenic, as well as other substances. His lexicon was a means for interpreting the variety of terms used for arsenic; especially those used by the Paracelsians.

Conclusion

Known from ancient times, the arsenic sulfides found their way into various uses as paint pigments, medicinal treatments, poisons, and substances used in the alchemists' investigations. The preeminent pursuits of the alchemists involved the discovery of ever-more effective medicines and the transmutation of base metals. In regard to the latter, the alchemists offered recipe upon recipe, each one failing to truly transmute a base metal into a more noble metal—such as the transmutation of copper or iron into silver. Success, however, was achieved in finding methods for tingeing metals such as imparting a silvery color to copper or iron and alloying copper or iron into stronger materials. Another possible achievement of the early alchemists was that of isolating metallic arsenic. This, of course, has been highly disputed, but the possibility that metallic arsenic was separated out from the arsenic minerals, even without recognizing elemental arsenic, remains.

Notes

1. Etymonline- Online Etymology Dictionary, "Arsenic," https://www.etymonline.com/word/arsenic. Retrieved 1 July 2023. The word arsenic had its origin in the Syriac word "zarnika," which was from the Arabic *al-zarnik*, the orpiment, based on the Persian word "zar" (gold) from the word "zarnikh," meaning yellow or gold-colored, hence yellow orpiment. It was adopted into Greek as *arsenikon*, which then was adopted in Latin, arsenicum, which in French became arsenic, from which the English word arsenic was taken.
2. Bronze is an alloy consisting primarily of copper, commonly with about 12–12.5% tin and often with the addition of other metals (including aluminum, manganese, nickel, or zinc) and sometimes non-metals, such as phosphorus, or metalloids such as arsenic or silicon. These additions produce a range of alloys that may be harder than copper alone or have other useful properties, such as strength, ductility, or machinability.
3. J. A. Charles, "Early Arsenical Bronzes—A Metallurgical View," *American Journal of Archaeology*, Vol. 71, No. 1 (Jan., 1967), pp. 21-26 (18 pages), Published By: The University of Chicago Press. See also, C.P. Thornton, Th. Rehren, V.C. Pigott, "The production of speiss (iron arsenide) during the Early Bronze Age in Iran," *Journal of Archaeological Science*, Volume 36, Issue 2, February 2009, Pages 308-31.
4. Arsenopyrite (FeAsS), Enargite (Cu_3AsS_4), Olivenite ($Cu_2(AsO_4)$OH), Tennantite ($Cu_{12}As_4S_{13}$), Malachite ($Cu_2(OH)2CO_3$), and Azurite ($Cu_3(OH)_2(CO_3)_2$), Lechtman & Klein, 1999. "The Production of Copper–Arsenic Alloys (Arsenic Bronze) by cosmelting: Modern Experiment, Ancient Practice," *Journal of Archaeological Science*. 26 (5): 497–526.
5. M. Harper, "Possible toxic metal exposure of prehistoric bronze workers," *British Journal of Industrial Medicine*, 1987; 44:652-65.
6. Philip Wheeler, *Alchemical Symbols, Fourth Edition, R.A.M.S. Library of Alchemy, Vol. 21*, (Kansas City, MO: R.A.M.S. Publishing, Co., 2018), 22.
7. The chemical formula of realgar is often written as As_4S_4 instead of the simpler AsS because As_4S_4 represents a structural unit of the mineral. A compound of As+3 and S-2 would be out of electrical balance. In realgar, three of the arsenics are joined in a chain by covalent bonds. This

gives the arsenics an effective electrical charge of +8, which combines with four S-2 ions to produce an electrically neutral molecule. This is why realgar's chemical composition is often presented as As_4S_4 instead of AsS.

8. Hobart M. King, PhD, RPG. "Realgar and Orpiment- Arsenic Sulfide Minerals," Geology.com. Retrieved 22 June 2023. See also John W. Anthony, Richard A. Bideaux, Kenneth W. Bladh, Monte C. Nichols, (2005), "Orpiment," Handbook of Mineralogy, (handbookofmineralogy.org).

9. D. L. Douglass, Chichang Shing, Ge Wang (1992). "The light-induced alteration of realgar to pararealgar," *American Mineralogist*, 77: 1266–1274. Retrieved 1 July 2023.

10. Wheeler, 60.

11. Online Etymology Dictionary, "Orpiment," https://www.etymonline.com/. Retrieved 1 July 2023.

12. Wheeler, 67.

13. Theophrastus (Ancient Greek: Θεόφραστος, romanized: *Theóphrastos*, lit. 'godly phrased'; c. 371 – c. 287 BC) was a Greek philosopher and the successor to Aristotle in the Peripatetic school. The principal interest in Theophrastus's "On Stones" is the description of minerals. He enumerates various exterior characteristics such as color, tenacity, hardness, smoothness, density, fusibility, luster, and transparency, and their quality of reproduction. It is possible to identify certain rocks and minerals, such as marble, pumice, onyx, gypsum, pyrites, coal, bitumen, amber, azurite, chrysocolla, realgar, orpiment, cinnabar, quartz in various forms, lapis lazuli, emerald, sapphire, diamond, and ruby (*De Re Metallica*, Appendix B).

14. A Roman city in ancient Paphlagonia located in the Kastamonu Province in the Black Sea Region of Turkey.

15. Nicander, *Alexipharmaca*. See also Plutarch's, *Themistocles*, " …and, having entertained them and shaken hands with them, drank bull's blood, as is the usual story; as others state, a poison producing instant death; and ended his days in the city of Magnesia…" The Internet Classics Archive, "Themistocles by Plutarch" (mit.edu). Retrieved 17 July 2023.

16. RRUFF Project, "Realgar," Handbook of Mineralogy. https://rruff.info/doclib/hom/realgar.pdf. (Retrieved 1 July 2023). See also King;

and List of Minerals, The International Mineralogical Association, https://mineralogy-ima.org/Minlist.htm. Retrieved 1 July 2023.
17. Matteo Martelli, *The Four Books of Pseudo-Democritus,* Sources of Alchemy and Chemistry, *Ambix* Volume 60, Supplement 1, 2013 (London: Routledge), S36-S44.
18. Martelli, referenced on contents page.
19. Martelli.
20. F. Sherwood Taylor, *Alchemists, Founders of Modern Chemistry,* Kessinger Legacy Reprints, (Kessinger Publishing), 32-33.
21. Martelli.
22. To the author's knowledge, these recipes of Pseudo-Democritus have not been proven through laboratory experimentation. Their interpretations are mostly theoretical, based on modern chemical knowledge.
23. Pliny the Elder, *Natura Historia.*
24. Based on Zosimos's writings, Maria lived between the first and third centuries CE in Alexandria.
25. H. E. Stapleton has postulated that Agathodaimon and Maria the Jewess lived in Seleucid Syria rather than Ptolemaic Egypt between the fourth and first centuries B.C., thus placing the genesis of alchemy from Ptolemaic Egypt to Seleucid Syria. Stapleton, "The Antiquity of Alchemy," *Ambix*, Vol. 5, Nos. 1 & 2, October 1953. See also Multhauf, *The Origins of Chemistry,* 104-105.
26. Maddalena Rumor and Matteo Martelli, "Near Eastern origins of Graeco-Egyptian Alchemy," Researchgate, (https://www.researchgate.net/publication/328809940). The *Anepigraphos* was possibly written around the 2nd or 3rd century CE (the author is unknown), cites Hermes and Agathodaemon as the authorities for an allegory for the making of silver, called the "moon," from various substances. See Brian P. Copenhaver, *Hermetica: The Greek Corpus Hermeticum and the Latin Asclepius,* (Cambridge: Cambridge University Press, 1992), xxxiv.
27. John Emsley, *The Elements of Murder: A History of Poison,* (Oxford: Oxford University Press, 2005), 117.
28. This may be a Hellenized form of the ancient Hittite Moon-God ARMA, cf. E.O. von Lippman, *Ambix*, 1938, II, p. 21.
29. H. E. Stapleton, "Summary of the Cairo Arabic MS of the Treatise of Agathodaimon: his discourse to his disciples when he was about to die," Appendix B., "The Antiquity of Alchemy," *Ambix*, Vol. V, Nos.1-2, 1-43.

30. Multhauf, Robert P., Smithsonian Institution Archives. https://siarchives.si.edu/collections/siris_arc_217620. Retrieved 3 July 2023.
31. A device used to heat substances and collect vapors invented by Maria the Jewess. She was also credited with the invention of several kinds of chemical apparatus, including the tribikos (a kind of alembic or retort), the kerotakis (a device used to heat substances and collect vapors), and the Bain-marie (Maria's bath, which limits the maximum temperature of the container and its contents to the boiling point of a separate liquid).
32. Robert P. Multhauf, *The Origins of Chemistry*, (Langhorne, PA: Gordon and Breach Science Publishers, 1993), 108.
33. Sherwood Taylor, F. (1937). "The Visions of Zosimos," *Ambix*, 1 (1): 88–92. doi:10.1179/amb.1937.1.1.88.
34. Alchemy at Clifton. https://web.archive.org/web/20070221092632/http://www.scienceatclifton.co.uk/clifton/alchemy.htm. Retrieved 3 July 2023.
35. According to *Liber de aluminibus et salibus*, the only witness to an Arabic dictionary of alchemical terms, "Bird" could refer to sal ammoniac, to arsenic, or to sulfur. In addition, "Lion" was also used to refer to arsenic as well as silver or sal ammoniac. See Gabriele Ferrario, "An Arabic Dictionary of Technical Alchemical Terms:" MS Sprenger 1908 of the Staatsbibliothek zu Berlin (fols. 3r–6r), *Ambix*, Vol. 56, No. 1, March 2009, 36–48.
36. John Eric Holmyard, *Makers of Chemistry* (Oxford: Oxford University Press, 1931). Makers of Chemistry : Holmyard, John Eric : Free Download, Borrow, and Streaming : Internet Archive.
37. Xingxing Wang et al., "Oxidative dissolution of orpiment and realgar induced by dissolved and solid Mn (III) species," *Geochimica et Cosmochimica Acta*, Vol. 332, 1 September 2022, 307–326.
38. Arsenious Oxide, Sublimation, Big Chemical Encyclopedia. Chempedia.info. Retrieved 3 July 2023.
39. Many metals, including magnesium, aluminum, zinc, tin, and iron, precipitate arsenic and liberate arsine (AsH_3) from aqueous arsenic acid (H_3AsO_4). When copper is placed in such a solution containing mineral acid, copper arsenide ($AsCu_3$) is formed on the metal. "Copper Arsenide," Big Chemical Encyclopedia, https://chempedia.info/page/119253014000119132173095176107202159007024139055/. Retrieved 5 July 2023.

40. George Sarton, *Introduction to the History of Science, Vol. 1: From Homer to Omar Khayyam*, (Baltimore: Carnegie Institution of Washington, 1927), 532. See also, Antoine-François Fourcroy,(1804), *A general system of chemical knowledge, and its application to the phenomena of nature and art*. 84
41. Harran, a district in southern Turkey, which was an important commercial center between the Tigris and the Euphrates.
42. A collection of possibly 600 writings (215 are extant today) attributed to Abū Mūsā Jābir ibn Ḥayyān, purported to have lived during the 9th century but whose actual existence has been questioned. The works that survive today mainly deal with alchemy and chemistry, magic, and Shi'ite religious philosophy. However, the original scope of the corpus covered a wide range of topics ranging from cosmology, astronomy, and astrology, over medicine, pharmacology, zoology, and botany, to metaphysics, logic, and grammar. The authenticity of the writings has also been questioned since much of the terminology used was not coined until later in the 9th and 10th centuries. See Newman, William R. (1985), "New Light on the Identity of Geber," Sudhoffs Archive 69 (1): 76–90. *JSTOR* 20776956. PMID 2932819.
43. Ana Maria Alfonso-Goldfarb & Safa Abou Chahla Jubran, "Listening to the Whispers of Matter through Arabic Hermeticism: New Studies on the Book of the Treasure of Alexander," *Ambix*, Vol. 55, No. 2, July 2008, 99–121.
44. Ibid.
45. "Frankincense," Wikipedia. Retrieved 7 July 2023.
46. "Incensole Acetate," NIST Chemistry Webbook, https://webbook.nist.gov/cgi/inchi?ID=C34701536&Mask=2000. Retrieved 7 July 2023.
47. Ana Maria Alfonso-Goldfarb & Safa Abou Chahla Jubran.
48. "Ratl," Wikipedia. A medieval Middle Eastern unit of measurement found in several historic recipes. The term was used to measure both liquid and solids (approximately a pint and a pound in 10th-century Baghdad, but could range from 8 ounces to 8 pounds depending on the time period and region). See al-Warrāq, al-Muẓaffar Ibn Naṣr Ibn Sayyār (2007-11-26), "Annals of the Caliphs' Kitchens: Ibn Sayyār Al-Warrāq's Tenth-Century Baghdadi Cookbook," *JSTOR*, Review: [Untitled] on JSTOR. Retrieved 7 July 2023.

49. Arabic (archaic)- possible reference to silver.
50. Ana Maria Alfonso-Goldfarb & Safa Abou Chahla Jubran.
51. E.J. Holmyard, *Alchemy*, (Garden City, NY: Dover, 1990), 49.
52. A technique for joining two different structures.
53. This may have been zinc. See H.E. Stapleton and R. F. Azo, and M. Hidayat Husain, "Chemistry in Iraq and Persia in the Tenth Century A.D.," *Memoirs of the Asiatic Society of Bengal*, 8 (1927): 321n3.
54. Marcasite—a metallic sulfide; Magnesia—manganese ore; Tutia—zinc ore; Lapis lazuli—azurite, a blue semi-precious stone composed chiefly of a sulfur-containing silicate of sodium and aluminum; Malachite—a monoclinic basic copper carbonate, usually occurring as bright-green masses of fibrous aggregates, which is used ornamentally and as an ore of copper; White arsenic—arsenic derived from smelting arsenic-laden silver; Kohl—a sulfide of antimony used in powdered form as an eye makeup; Talc—mica, and transparent plates that split off some varieties of gypsum; Gypsum—hydrated calcium sulfate, a soft mineral that occurs as colorless, white, or grey monoclinic prismatic crystals in many sedentary rocks and is used in making plaster of Paris and fertilizer. See Gail Marlow Taylor, *The Alchemy of Al-Razi, A Translation of the "Book of Secrets,"* (North Charleston, South Carolina: Createspace Independent Publishing Platform, 2014), 102n18.
55. Gail Marlow Taylor, *The Alchemy of Al-Razi: A Translation of the "Book of Secrets,"* (North Charleston, South Carolina: Createspace Independent Publishing Platform, 2014), 103-104, 104n27.
56. Darrell Kirk Nordstrom and Reza Sharifi, "Recent Studies of arsenic mineralization in Iran and its effect on water and human health," Researchgate.net. Retrieved 11 July 2023.
57. Encyclopedia Iranica, "Mining in Iran: Mines and Mineral Resources," https://www.iranicaonline.org/articles/mining-in-iran-i. Retrieved 11 July 2023.
58. "Olivenite and Lammerite," Handbook of Mineralogy, Mindat.org. https://rruff.info/doclib/hom/olivenite. Retrieved 11 July 2023.
59. Pierfranco Lattanzi et al., "Enargite oxidation: A review," *Earth-Science Reviews*, Volume 86, Issues 1-4, January 2008, Pages 62-88. Science Direct (sciencedirect.com). Retrieved 12 July 2023. See also, Yu Zhao, Hongbo Zhao, Tatiana Abashina, and Mikhail Vainshtein, "Review on arsenic removal from sulfide minerals: An emphasis on enargite

and arsenopyrite," *Minerals Engineering*, Volume 172, 1 October 2021, 107133. Science Direct, sciencedirect.com. Retrieved 12 July 2023.
60. Taylor, 101-102, 102n15, 106.
61. Yu Zhao, Hongbo Zhao, Tatiana Abashina, and Mikhail Vainshtein, "Review on arsenic removal from sulfide minerals: An emphasis on enargite and arsenopyrite," *Minerals Engineering*, Volume 172, 1 October 2021. Science Direct, sciencedirect.com. Retrieved 12 July 2023.
62. Taylor, 139n85, n86. Sea foam—zubd al-bahr or zabad al-bahr may be understood as butter or sea foam and could possibly mean foamy borax. Mashaquiniya—a salt-like drying substance used in glass-making.
63. Taylor, 139.
64. Ibid, 139-140.
65. Alan Williams, "A Note on Liquid Iron in Medieval Europe," *Ambix*, Vol. 56 No. 1, March, 2009, 68–75.
66. Clifford A. Pickover, *The Math Book: From Pythagoras to the 57th Dimension, 250 Milestones in the History of Mathematics*, (Sterling Publishing Company, Inc., 2009), 84. See also Isaac Asimov, *Asimov's Biographical Encyclopedia of Science & Technology*, (Garden City, New York: Doubleday, 1982), 79.
67. Samanid Nuh II ibn Mansur ruled over Transoxania and Khorasan. Transoxania was the name for a region and civilization located in lower Central Asia roughly corresponding to modern-day eastern Uzbekistan, western Tajikistan, parts of southern Kazakhstan, parts of Turkmenistan, and southern Kyrgyzstan. Khorasan comprised the present territories of northeastern Iran, parts of Afghanistan, and southern parts of Central Asia. *Encyclopedia Britannica*. Retrieved 7 July 2023.
68. Karin C. Ryding, " Islamic Alchemy According to Al-Khwarizmi," *Ambix*, Vol. 41, Part 3, November 1994.
69. Ibid.
70. Stanton J. Linden, editor, *The Alchemy Reader: From Hermes Trismegistus to Isaac Newton* (Cambridge: Cambridge University Press, 2003), 104–105.
71. Vladimír Karpenko, "Not All That Glitters is Gold: Gold Imitations in History," *Ambix*, Vol. 54, No. 2, July 2007, 172–191.
72. *Webster's Unabridged Dictionary of the English Language*, "Imbibition." The absorption of a liquid. In physical chemistry, the absorption of a solvent by a gel.

73. *Webster's Unabridged Dictionary of the English Language*, "Ampulla." A two-handled bottle having a somewhat globular shape, made of glass or earthenware.
74. Linden, 110.
75. "Arsenic," Encyclopedia Britannica Online. Retrieved 16 July 2023.
76. "Potash," Wikipedia, https://en.wikipedia.org/wiki/Potash. Retrieved 18 July 2023.
77. John Emsley, *Nature's Building Blocks, An A-Z Guide to the Elements*, (Oxford: Oxford University Press, 2011), 47-55. See also Encyclopedia Britannica Online, "Arsenic."
78. Eugene Garfield, "George Sarton: The Father of the History of Science. Part 1.Sarton's Early Life in Belgium," *Essays of an Information Scientist*, Vol. 8, p.241-247, 1985, Current Contents, #25, p.3-9, June 24, 1985. Institute for Scientific Information. See also George Sarton, *Introduction to the History of Science*.
79. Wheeler, 22.
80. Harold J. Abrahams, "Al-Jawbari on False Alchemists," *Ambix*, Vol. 31, Part 2, July 1984.
81. Geber, *Summa Perfectionis*, The R.A.M.S. *Library of Alchemy, Vol. 9*, (Stuarts Draft, VA: R.A.M.S. Publishing, 2015), 25.
82. E.J. Holmyard, *The Works of Geber*, (Kessinger Publishing, 1928), 210.
83. Dictionary of Gnosis & Western Esotericism, https://referenceworks.brillonline.com/entries/dictionary-of-gnosis-and-western-esotiricism/sedacer-guillaume-DGWE_342. Retrieved 12 July 2023.
84. May refer to plums.
85. Alan Williams, "A Note on Liquid Iron in Medieval Europe," *Ambix*, Vol. 56 No. 1, March, 2009, 68–75.
86. Paracelsus, *De Naturalibus Rebus*. See *The Hermetic and Alchemical Writings of Aureolis Philippus Theophrastus Bombast, of Hohenheim, Called Paracelsus the Great*. Edited by Arthur Edward Waite in two volumes, (Mansfield Centre, CT: Martino Publishing, 2009). 59n.
87. Paracelsus, *The Hermetic and Alchemical Writings of Aureolis Philippus Theophrastus Bombast, of Hohenheim, Called Paracelsus the Great*. Edited by Arthur Edward Waite in two volumes, (Mansfield Centre, CT: Martino Publishing, 2009). 59-60.
88. Paracelsus and Waite, 44, 78.
89. To make a volatile subject fixed or solid, so that it remains permanently

unaffected by fire. The Alchemy Website, https://alchemywebsite.com. Retrieved 13 July 2023.
90. A crucible or cupel.
91. The act or process of boiling up. *Webster's Unabridged Dictionary of the English Language*, (New York: Random House, 2001).
92. Paracelsus, *The Hermetic and Alchemical Writings of Aureolis Philippus Theophrastus Bombast, of Hohenheim, Called Paracelsus the Great*. Edited by Arthur Edward Waite in two volumes, (Mansfield Centre, CT: Martino Publishing, 2009). 58n. Also in Chirurgia Minor, Lib. II. Again: The fixation of arsenic is performed by salt of urine, after which it is converted by itself into an oil.
93. Paracelsus, Ibid.
94. The identities of Aristotle the Chemist or Polydorus are unclear. Rhasis, another name for al-Razi.
95. Paracelsus, *The Hermetic and Alchemical Writings*, 59.
96. Georgius Agricola, *De Natura Fossilium (A Textbook of Mineralogy)*, Translated from the first Latin Edition of 1546 by Mark Chance Bandy and Jean A. Bandy (Mineola, NY: Dover Publications, Inc., 2004), 202.
97. A Roman architect and engineer during the 1st century BC, known for his multi-volume work titled *De architectura*.
98. A region in the northwest of ancient Asia Minor (Anatolia, Asian part of modern Turkey).
99. A region on the southern coast of the Black Sea, located in the modern-day eastern Black Sea Region of Turkey.
100. A region in Central Anatolia, Turkey.
101. Agricola, *De Natura Fossilium (A Textbook of Mineralogy)*.
102. A copper mineral, hydrous copper phyllosilicate.
103. A highly viscous liquid or semi-solid form of petroleum.
104. Georgius Agricola, *De Re Metallica*, translated from the first Latin edition of 1556 by Herbert Clark Hoover and Lou Henry Hoover (New York: Dover Publications, Inc., 1950), 235.
105. Georgius Agricola, *De Natura Fossilium (A Textbook of Mineralogy)*, 58.
106. Agricola, *De Re Metallica*, 111n.
107. Agricola, *De Natura Fossilium (A Textbook of Mineralogy)*, 57-58.
108. Agricola, *De Re Metallica*, 113, 214n21.

109. Spirit of Turpentine.
110. Martinus Rulandus, *A Lexicon of Alchemy, Containing a full and plain explanation of all obscure words, hermetic subjects, and arcane phrases of Paracelsus* (Zachariah Palthenus, bookseller, in the free republic of Frankfurt, 1612), 49, 274-275, 322.

Chapter 10

The Vitriols

Introduction

The mineral substances known as vitriols throughout antiquity and the Middle Ages, and up to the Modern Age (Vitriol was an archaic name for sulfate),[1] are today understood to be the hydrated sulfates of primarily iron and copper, but of zinc and possibly of aluminum, as well. (Though aluminum sulfate was most likely potassium aluminum sulfate, and referred to as alum or Alom.)[2] These substances formed as secondary minerals in the weathered metallic sulfide deposits, which were referred to as pyrites during antiquity.[3]

The vitriols were probably discovered during the mining of the copper and iron sulfide deposits, which would have formed from the weathering of such deposits. The presence of vitriols, however, would have been fleeting since these sulfates were highly soluble and prone to degradation by absorbing water. Thus, the vitriols used commercially or in the laboratories were probably extracted from solutions of decomposing sulfide and sulfate minerals.[4]

The alchemists frequently classified the vitriols by color, though other classifications were devised, such as the one used by Georgius Agricola (16th-century metallurgist) to distinguish between artificial and natural formation of the vitriols or by location, such as those found in the listings of Martin Rulandus's lexicon (see below). The color schemes identified several types of vitriol—blue, green, and white.

Blue Vitriol

Blue vitriol, now known as copper sulfate ($CuSO_4$), corresponded to the modern mineralogical term chalcanthite—pentahydrate copper sulfate

($CuSO_4 \cdot 5H_2O$)—a blue-green, water-soluble mineral commonly found in the late-stage oxidation zones of copper deposits. Since chalcanthite is readily soluble in water, it tends to crystallize, dissolve, and recrystallize as crusts over mine surfaces in more humid areas. Therefore, undissolved chalcanthite is only found in arid regions[5] in sufficiently large quantities for use as an ore.

Blue Vitriol

Given that the chalcanthite[6] group is found in oxidized copper deposits, it is frequently found in association with other copper minerals, which include: calcite and its polymorph, aragonite, both $CaCO_3$; brochantite, $Cu_4(SO_4)(OH)_6$; chalcopyrite, $CuFeS_2$; malachite, $Cu_2(CO_3)(OH)_2$; and melanterite, $FeSO_4 \cdot 7H_2O$. The ores involved, upon exposure to air, oxidize to a variety of oxides, hydroxides, and sulfates.

Green Vitriol
Known since ancient times as copperas, Roman Vitriol, or green vitriol, the blue-green heptahydrate (hydrate with 7 water molecules) is the most common form of iron (II) sulfate $FeSO_4 \cdot 7H_2O$ or ferrous sulfate. All the iron(II) sulfates dissolve in water to give the same aquo complex $[Fe(H_2O)_6]^{2+}$, which has octahedral molecular geometry and is paramagnetic. The name copperas dates from times when the copper (II) sulfate was known as blue copperas, and perhaps in analogy, iron(II) and zinc sulfate were known respectively as green and white copperas.[7] Ferrous sulfate was used in the manufacture of inks, most notably iron gall ink, which was used from the Middle Ages until the end of the 18th century. Blackening was a distinguishing property of iron sulfate when used in dyeing or as an ink, and it was often a means for determining the type of vitriol. Today, the hydrated form is used medically to treat iron deficiency, and it also has many industrial applications.

Green Vitriol

Blue and Green Vitriol Confusion

The alchemists did not differentiate the vitriols in regard to actual composition. (The ability to ascertain the specific composition was not yet available.) So color and method of formation were the primary differentiating factors, and sometimes the coloring shaded from a bluish-green to green. Thus, vitriol was often given names associated with copper while its use often indicated the presence of iron sulfate. For instance, because of vitriol's association with sulfide ores that were mined for copper, it was commonly thought to have a cupriferous nature, which led the Greeks to call it chalcanthon, meaning "copper flower." In Latin, it was called *Atramentum sutorium* (tanner's ink),[8] referring to its principal commercial use as a blackening agent for leather. This ability to blacken leather, however, was a property of iron sulfate, and, given that the composition of the sulfide deposits on Cyprus was actually iron-rich, the vitriol most likely contained iron sulfate.[9]

Atramentum

White Vitriol

Zinc sulfate describes a family of inorganic compounds with the formula $ZnSO_4(H_2O)x$. All are colorless solids, the most common form of which includes the heptahydrate, with the formula $ZnSO_4 \cdot 7H_2O$. It was historically known as "white vitriol."

White Vitriol

Alum

Alum was a special case among the vitriols (further discussed below). Among the ancients and the medieval alchemists, alum was most likely aluminum sulfate. A salt with the formula $Al_2(SO_4)_3$, aluminum sulfate is soluble in water and occurs naturally in the anhydrous form as the rare mineral millosevichite, which is found, for example, in volcanic environments and on burning coal-mining waste dumps. Aluminum sulfate is rarely encountered as the anhydrous salt, however. It forms a number of different hydrates, of which the hexadecahydrate $Al_2(SO_4)_3 \cdot 16H_2O$ and octadecahydrate $Al_2(SO_4)_3 \cdot 18H_2O$ are the most common. The heptadecahydrate, $[Al(H_2O)_6]_2(SO_4)_3 \cdot 5H_2O$, occurs naturally as the mineral alunogen.[10] In alchemical writings, the term "Alum" or "Allom" was used instead of terms typically used for the vitriols. Nevertheless, alum was included in the listings of vitriols (For example, see the writings of the Persian alchemist al-Razi, discussed below).

Alum

Vitriol in Alchemical Literature
Dioscorides, Pseudo-Democritus, and Pliny the Elder

The earliest known and most reliable descriptions of vitriol in Europe were found in the works of the Greek physician Dioscorides (c. 40-90), Pseudo-Democritus (First Century CE), and the Roman naturalist Pliny the Elder (c. 23-79). Most of Dioscorides's work was oriented toward herbs and medical matters. However, he devoted Book V of his *De Materia Medica* to minerals and ores and their preparation for medicinal purposes. In addition to vitriol, which he referred to as calkanthon,[11] the work identified several related substances, including chalcitis (copper sulfide as either CuS or CuS_2), misy, melanteria, and sory (copper sulfide—CuS or CuS_2—or iron sulfide—FeS or Fe_2S_3—oxidation minerals), and alum. Copper sulfide was also known as *chalcocite*,[12] vitreous copper or copper-glance.

5-114. Chalkanthon
There is a single type of calcanthum formed by moisture into a solid. It has three different forms. The moisture that is strained by dripping into certain caves is formed into a solid from which those who work the [Cyprian] metals call stillatitium. Petesius calls it pinarium, and some call it stalacticon [coalesced, boiled, thin, long like a spear]. Some runs secretly in hollows, and afterwards is transported into ditches and thickens, and this is called pecton.

The third sort is called coctile and is made in Spain. It has the following method of manufacture, but is useless and the weakest. They dilute it in boiling water, and then pour it out into receptacles to let it stand. After some days, this is thickened and divides into many cubic forms hanging together like grape bunches. The best is azure color [blue], heavy, compact, and transparent, such as the stillatitium that is also called lanceatum. The next best is concretitium. Coctile is thought to be the fittest for dyeing and making colors black, but experience shows that it is the weakest for medicinal use. It is astringent — two teaspoonfuls are swallowed or licked with honey to warm, form scabs, and to kill broadworms. It causes vomiting and helps those who have eaten mushrooms, taken as a drink with water. Diluted with water and dropped into the nostrils through wool, it purges the head. It is burnt, as we will show [below] in calcitis.[13]

Though Dioscorides's primary purpose was to describe the use of vitriol (*calcanthum*) for medical treatments, his description contained significant information regarding the formation and the types of vitriol. First, *calcanthum* was seen as one type having three different forms which were solidified liquids, formed by the addition of moisture and differentiated by the manner of their formation. One was formed by the liquids which trickled down and congealed in the mines. A second type formed in caves or hollows. A third, called *coctile* (Gk. i.e., cooked), was manufactured in Spain.[14]

First discussed by Dioscorides, the related substances—chalcitis, misy, sory, and melanteria–were significant to the discussion of vitriols because of their use in manufacturing vitriols (For example, see Georgius Agricola's *De Re Metallica* discussed below).

5-115. Chalkitis
Calcite—Anhydrous Carbonate of Lime, Calcium Carbonate
Chalcitis is preferred, which looks like brass, brittle, without stones, and is not old, and furthermore, with somewhat long glittering veins. It is astringent, warming, and scab-forming...

5-117. Misy
Misy[15]—Copiapite[16]
Cyprian Misy must be chosen—looking like gold, hard, and if broken, a golden color, glistening like a star. It has the same strength as Chalcitis and is burnt the same way without psoricum[17] being produced from it, differing both in excess and defect. That from Egypt (compared to others) is the best and most effective, but for eye medicines, it is not comparable to that previously mentioned.

5-118. Melantheria
Melantheria grows together in the manner of salt at the mouths of mines out of which brass is dug, and some comes from the earthy upper surfaces of these places. Some is also found that is dug out in Cilicia and in certain other places. The best is a sulphurous color [yellow], smooth, even, and clear, and on touching water, it quickly turns black. It is sharp like Misy [above].

5-119. Soru [Sory]
Some are deceived, supposing Sory to be Melanteria [above], for it is a different thing though not unlike. Sory is more poisonous and causes nausea. It is found in Egypt and in other places such as Africa, Spain, and Cyprus. The Egyptian is considered the best—that which looks blackest when broken, has many holes, is somewhat fat, and also astringent and poisonous to taste or smell, overturning the stomach. That which does not glisten when broken (like Misy) is thought to be another kind and weak. It has the same properties as those mentioned above and is burnt like them... It is mixed [with other ingredients] for dyeing hair black. As a general rule, for these and almost all others, those which are not burnt are considered to be stronger than the burnt; except for salt, wine sediment, saltpeter [potassium nitrate], chalk, and other similar things which are weak raw, but are more effective burnt.[18]

Clues that sory was a sulfide, or had decomposed to a sulfate, were its blackening ability and its astringent taste and smell. As iron (II) sulfide, it would easily oxidize in moist air to become hydrated ferrous sulfate (FeS + H_2SO_4 → $FeSO_4$ + H_2S), and as iron (III) sulfide it would have decomposed easily above 20°C into iron (II) sulfide and elemental sulfur.

In addition, Dioscorides discussed alum (Stupteria) in *De Materia Medica*, Book V, chapter 123, in which he emphasized the astringent taste and the odor of fire, which suggested a sulfate:

> 5-123. *Stupteria*
> *Aluminum Sulfate, Potassium Sulfate—Alum*
> Almost every kind of Allom is found in the same mines in Egypt, such as the scissile [capable of being cut or divided] (as it were) and the flower of Bolitis. It is also found in certain other places—in Melos, Macedonia, Sardinia, Liparis, and Hierapolis in Phrygia, in Africa, Armenia, and many other places (like red ochre). There are many kinds of it; but for medicinal use the scissile, the round, and the moist are taken. The scissile is the best—especially that which is new, white, without stones, with a strong smell, very astringent, and furthermore not compacted together like turf or slate, but opening its mouth wide like gray hairs spread-apart, such as that called trichitis, found in Egypt. A stone very like it is also found, discernible by its not astringent taste . . . Moist [Alum] that is most transparent must be chosen—milky, even, and juicy throughout, and furthermore without stones and giving out a smell of fire. It is warming, astringent. . . .[19]

Pseudo-Democritus

Pseudo-Democritus was also aware of the mineral of sory. For instance in his "On the Making of Purple and Gold: Natural and Secret Questions," he recorded a recipe using sory and copper flower in either coloring or transmuting silver to gold.

> Grind sori and copper flower together with unburnt sulphur; sori looks like azurite that easily peels off and is always found in misy. It is also called green copper flower. Roast it at moderate heat for three days, until it becomes a yellow drug. Lay it on the copper or on the silver that is made by us and it will be gold.[20]

CHAPTER 10

Pliny the Elder

Pliny the Elder, in Book 34 of his work *Natural History,* tied the Greek term for vitriol, *calcanthum,* to the Latin term *Atramentum sutorium* (tanner's ink), and although his description contained much of the same information as Dioscorides's (Pliny may have known Dioscorides's work), Pliny distinguished vitriol by two kinds, further differentiated by their color:

> *The Greeks, by the name which they have given to it, have indicated the relation between shoemakers' black and copper; for they call it "chalcanthum." Indeed, there is no substance so singular in its nature. It is prepared in Spain, from the water of wells or pits which contain it in dissolution. This water is boiled with an equal quantity of pure water, and is then poured into large wooden reservoirs. Across these reservoirs, there are a number of immovable beams, to which cords are fastened, and then sunk into the water beneath by means of stones; upon which, a slimy sediment attaches itself to the cords, in drops of a vitreous appearance, somewhat resembling a bunch of grapes. Upon being removed, it is dried for thirty days. It is of an azure color, and of a brilliant luster, and is often taken for glass. When dissolved, it forms the black dye that is used for coloring leather.*
>
> *Chalcanthum is also prepared in various other ways: the earth which contains it being sometimes excavated into trenches, from the sides of which globules exude, which become concrete when exposed to the action of the winter frosts. This kind is called "stalagmia," and there is none more pure. When its color is nearly white, with a slight tinge of violet, it is called "lonchoton." It is also prepared in pans hollowed out in the rocks; the rainwater carrying the slime into them, where it settles and becomes hardened. It is also formed in the same way in which we prepare salt; the intense heat of the sun separating the fresh water from it. Hence it is that some distinguish two kinds of chalcanthum, the fossil and the artificial; the latter being paler than the former, and as much inferior to it in quality as it is in color.*[21]

Atramentum sutorium was of two kinds, according to Pliny: the fossil, or mineral, and the artificial—the latter paler than, and inferior to, the former. (Presumably, this was a reference to the artificial form's effectiveness in its applications). Pliny focused on the manufacturing method more than Dioscorides did, and on the prominence of blue vitriol. Pliny's solution for coloring leather must have been iron sulfate, however, since it produced the blackening, while the copper sulfate did not have that effect. Thus the method as described by Pliny would have resulted in iron vitriol.[22] In addition, comparing Pliny's description to the work of Agricola (see below), the "cords" that were sunk into the reservoirs were most likely made of iron which would have caused any copper to precipitate out, yielding iron sulfate.

Moreover, Pliny recognized the presence of the associated minerals in the copper-bearing ore (known as the chalcanthite group). In the chapter of his Book 34, *Copper and Bronze Sculpture; Tin, Lead, and Iron,* he described what he called copper-stone. "Copper-stone further differs from cadmea[23] in that it contains three kinds of minerals—copper, misy [copper pyrites], and sory [marcasite]."[24] Following Dioscorides, Pliny also provided descriptions of misy, sory, alum, and chalcitis from which copper was extracted:

Marcasite

Cadmia or Zinc Oxide

> CHAP. 29.—*Chalcitis: seven remedies.*
> *Chalcitis is the name of a mineral from which, as well as cadmia, copper is extracted by heat. It differs from cadmia in this respect that this last is procured from beds below the surface, while Chalcitis is detached from rocks that are exposed to the air. Chalcitis also becomes*

immediately friable, being naturally so soft as to have the appearance of a compressed mass of down. There is also this other distinction between them, that Chalcitis is a composition of three other substances, copper, Misy, and Sory, of which last we shall speak in their appropriate places. The veins of copper which it contains are oblong. The most approved kind is of the color of honey; it is streaked with fine sinuous veins, and is friable and not stony. It is generally thought to be most valuable when fresh, as when old, it becomes converted into Sory.

CHAP. 30.—Sory: three remedies.
The Sory of Egypt is the most esteemed, being considered much superior to that of Cyprus, Spain, and Africa; although some prefer the Sory from Cyprus for affections of the eyes. But from whatever place it comes, the best is that which has the strongest odor, and which, when triturated, becomes greasy, black, and spongy. It is a substance so unpleasant to the stomach that some persons are made sick merely by its smell. This is the case more particularly with the Sory from Egypt. That from other countries, by trituration, acquires the luster of Misy, and is of a more gritty consistency. Held in the mouth, and used as a collutory,[25] it is good for toothache. It is also useful for malignant ulcers of a serpiginous nature. It is calcined upon charcoal, like Chalcitis.

CHAP. 31.—Misy: thirteen remedies.
Some persons have stated that Misy is formed by the calcination of the mineral in trenches; its fine yellow powder becoming mixed with the ashes of the burnt firewood. The fact is, however, that though obtained from the mineral, it is already formed and in compact masses, which require force to detach them. The best is that which comes from the manufactories of Cyprus, its characteristics being that when broken, it sparkles like gold, and when triturated, it presents a sandy or earthy appearance, like Chalcitis. Misy is used in the process of refining gold.[26]

Pliny also wrote a description of alum, which was similar to Dioscorides':

In Cyprus, there is a white alum and another sort of a darker color, though the difference of color is only slight; nevertheless, the use made of them is very different, as the white and liquid kind is most useful for dyeing woolens a bright color, whereas the black kind is best for dark or somber hues. Black alum is also used in cleaning gold. All alum is produced from water and slime, that is, a substance exuded by the earth; this collects naturally in a hollow in winter, and its maturity by crystallization is completed by the sunshine of summer; the part of it that separates earliest is whiter in color. It occurs in Spain, Egypt, Armenia, Macedonia, Pontus, Africa, and the islands Sardinia, Melos, Lipara, and Strongyle; the most highly valued is in Egypt, and the next best in Melos. The alum of Melos also is of two kinds, fluid and dense. The test of the fluid kind is that it should be of a limpid, milky consistency, free from grit when rubbed between the fingers, and giving a slight glow of color; this kind is called in Greek 'phorimon' in the sense of 'abundant.' Its adulteration can be detected by means of the juice of a pomegranate, as this mixed with it does not turn it black if it is pure. The other kind is the pale rough alum which may be stained with oak-gall also, and consequently, this is called 'paraphoron,' perverted or adulterated alum. Liquid alum has an astringent, hardening, and corrosive property. Mixed with honey, it cures ulcers in the mouth, pimples, and eruptions; this treatment is carried out in baths containing two parts of honey to one of alum. It reduces odor from the armpits and perspiration. It is taken in pills against disorders of the spleen and discharge of blood in the urine. Mixed with soda and chamomile, it is also a remedy for scabies.

One kind of solid Alum, which is called in Greek schiston, 'splittable,' splits into a sort of filament of a whitish color, owing to which some people have preferred to give it in Greek the name of trichitis, 'hairy Alum' [having a fibrous nature]. This is produced from the same ore as copper, known as copper-stone, a sort of sweat from that mineral, coagulated into foam.[27]

According to Pliny, alum was found naturally in the earth, and the name distinguished different substances—phorimon, paraphoron, schiston, trichitis—all characterized by a certain degree of astringency and employed in dyeing and medicine. The light-colored alum was useful in brilliant dyes, while the dark-colored was used only in dyeing black or very dark colors. One kind of alum was a liquid, which was prone to be corrupted; but when purified, it had the property of blackening when it was added to pomegranate juice. This property indicated the presence of iron sulfate, since the blackening was more characteristic of an iron sulfate solution in water. A solution of ordinary (potash) alum or potassium aluminum sulfate would not possess such a property. Pliny also stated that there was another kind of alum (the Greeks called it *schistos*), which formed white threads on the surface of certain stones—an indication of alum slate and bituminous shale, consisting of iron and aluminum sulfates.[28]

Al-Razi (Abu Bakr Muhammad ibn Zakariya al-Razi c. 865-923)

In his *Book of Secrets,* al-Razi identified various vitriols by their color. He described two variations of yellow vitriol which he called *qalqatar*—one that was dense and hard and another yellow with "golden eyes"—in addition to *qalqadis* (white vitriol), *qalqant* (green vitriol), and *surin* (red vitriol):

> *The qalqadis is white vitriol, the qalqant is green vitriol, and the surin is red vitriol, and the qalqatar is yellow vitriol. These four are strong, and the surin is the strongest. It finds an application in the chapter on reddening. It is brought from the mines on Cyprus.*[29]

According to Vladimir Karpenko (Faculty of Science, Department of Physical and Macromolecular Chemistry, Charles University, Prague, Czech Republic) and John Norris (Ph.D. in History and Philosophy of Science, Charles University, Prague, Czech Republic), al-Razi's names for the vitriols corresponded to Arabic transliterations of the Greek: chalcathon, chalcitis, colcothar, and sory, which, as described by Dioscorides and Pliny, were derivatives of copper and iron sulfates.[30]

In addition, al-Razi included alum in his chapter on vitriols, listing various kinds as well: the white alum from Yemen, the Syrian (also a white alum), the Mingani (green alum), the Egyptian (a yellow alum), and another, unnamed, white form of alum:

> *Among them is the kind from Yemen, white (G. fibrous, fissile); in addition, the Syrian, white mixed with clay and rocks; then the Mingani, the color of which resembles green, then the yellow Egyptian and the white, stable kind.*[31]

Albertus Magnus (died 1280)

Unlike the terms *calcanthum* and *Atramentum sutorium*, the term vitriol did not appear in the literature until Albertus Magnus used it in his *De Mineralibus, Liber V*:[32]

> *The nature of Atramentum is peculiar to itself, a homeomerous [Being the same whether whole or divided into parts] mineral substance that can be dissolved by boiling in water, mixed with a stony substance that is not dissolved at all, even by strong boiling. The original kind of Atramentum is undoubtedly liquid, and it has solidified of its own accord. But all Atramentum, according to its kind, is characterized by having a foul taste, and in being styptic and very irritating; and therefore when it is applied to things, it thickens and hardens them.*
>
> *There are several forms: one kind is white, which the Arabs call alkadidis; one red, which they call assurie; and one yellow, which they call alkofol [colcothar]; and there is a green one which they call alkacantum; and a greyish one tinged with black which is almost stony. The green [kind], which some people call Vitriol (vitreolum) and classify as a sort of ink, is more firmly solidified than the yellow, and has thicker outside coatings. And the most efficacious among the greyish [kinds] of Atramentum is that showing a sort of golden glint, as if there were gold dust sprinkled through it and dimly gleaming.*
>
> *It is plain that almost all kinds of Atramentum are made of Earth and Water. At first, they were liquid and afterwards solid, and still, they can be re-dissolved by heat and moisture. Their [color] depends to a greater or lesser degree on the fine division of the earthy material in them and the thorough cooking of the moisture and on the larger or smaller amount of Air that is mixed in when the Earth was being cooked in the Water.*
>
> *[Atramentum] is thus an intermediate between stones and metals because it has the constitution of stone and sometimes the luster of metals.*[33]

The green kind, called *alkacantum*, was most certainly iron sulfate. Alkadidis (Arabic *qalqadis*, from Greek *chaldtis*) was a white or yellowish crust of oxidation products, most likely sulfates. Assurie (Arabic *al-suri*, corresponded to al-Razi's *surin* or the sory of Dioscorides and Pliny), was a red oxide of iron.

Alkofol was evidently the same word for alcohol, which originally meant "finely divided" or "subtle," but here it may have been an error for *colcothar* (Arabic *qalqatar*)—also known as Crocus Martis (yellow of Mars) which was a yellow iron oxide. The nameless blackish-gray material with a glint like gold dust was probably misy, containing undecomposed bits of pyrite or other sulfide minerals (Cf. Pliny, Nat. Hist. XXXIV, 31, above, in which he described misy as yellow and sparkling like gold.). But the substance to which *Atramentum* was more strictly applied was *alkacantum* (Arabic *qalqant*, from Greek *chalcanthon*), which included hydrous sulfates. [34]

Crocus Martis

Pseudo-Geber (1290-1330)
Pseudo-Geber used the locational term, Cyprus Vitriol, for vitriol in his recipe for making nitric acid:

> *Take a pound of Cyprus Vitriol [Fe,CuSO₄], a pound and a half of saltpeter, and a quarter of a pound of alum. Submit the whole to distillation, in order to withdraw a liquor which has a high solvent action. The dissolving power of the acid is greatly augmented if it be mixed with some sal ammoniac, for then it will dissolve gold, silver, and sulfur.*[35]

In another chapter, he identified vitriol as copperas, which was iron sulfate as evidenced by the red coloring (caused by the presence of iron oxide) of the acid produced by the recipe involved:

> *But the way of coloring it, which is made by vitriol, or copperas, is thus: ℞ of either of them a certain quantity, and sublime as much thereof, as can be sublimed...This being done, it is dissolved into a most red water....*[36]

Yet, other references in *Summa Perfectionis* simply listed vitriol, without further specific descriptions, along with nitre and alum in his recipes for *Aqua fortis*. "Take Luna 1 pound dissolved in its own water (made of nitre and vitriol)...dissolve it in *Aqua fortis* made of vitriol, nitre, and alum...."[37]

Al-Jildaki (died 1342)

Al-Jildaki's (an Islamic scholar and alchemist, born in Egypt) treatise *Kitab al-Burhan fi asrar 'ilm al-mizan*, provided a further understanding of the Arabic alchemy's terms for the vitriols and critiqued al-Razi's classification. Al-Jildaki asserted that there were seven kinds of vitriol—the yellow, the green, the red, qalqatar, qalqand, qalqadis, and shashira:

> ... *they are all natural-occurring minerals and from any one of them all the others may be prepared. Qalqadis is white vitriol, qalqand yellow, shading off into green and black, qalqatar is yellow vitriol in which are shining golden eyes. Al-suri is red vitriol; al-shashira is yellow shading off into green and pure blue. Al-Razi says that there are seven sorts of vitriol but mentions only six, and one of these is alum, which is not a vitriol; yet in another place he groups the alums and the vitriols separately in spite of what he said previously. In still another place he says that qalqadis is a white vitriol, which is correct, and that qalqand is green vitriol, but this is only partially true for it implies that every green vitriol is qalqand, and this is not so.*[38]

The uncertainty over how alum ought to be classified was apparent in al-Jildaki's disagreement with al-Razi on this issue. This disagreement may be understandable given that the sulfur may have been replaced by another chalcogen, or, more likely, the term was applied indiscriminately to various substances. For instance, it was applied to salty minerals which could contain a mixture of aluminum and iron sulfates. Also, according to Pliny, alumen was applied to any substance characterized by a certain degree of astringency.[39] In addition to the critique of al-Razi, al-Jildaki provided a more discriminatory description of the possible colors that could occur for specific vitriols. For instance, *qalqand*, which was yellow, could also be green or black. He also verified that *qalqatar* was *colcothar* ("yellow vitriol in which are shining golden eyes").

Basil Valentine (15th century)

In the work *From the Great Stone of the Ancient Sages,* attributed to Basil Valentine (The author published under this name may have been Johann Thölde, 1565—1614, the owner of a salt-works in Thuringia, Germany), the following reference to vitriol was found:

> *If you get such a deeply graduated and well-prepared mineral, called Vitriol [FeSO4], . . . put it into a well-coated retort, drive it gently at first, then increase the fire. There comes in the form of a white spirit of Vitriol [SO3] in the manner of a horrid fume, or wind, and cometh into the receiver as long as it hath any material in it . . . if you separate and free this expelled spirit well and purely per modum distillationis, from its earthy humidity [H2O], then in the bottom of the glass you will find the treasure, and fundamentals of all the Philosophers, and yet known to few, which is a red Oil, as ponderous in weight, as ever any Lead, or Gold may be, as thick as blood, of a burnt fiery quality.*

According to Karpenko and Norris, the viscosity and color of the "Oil" produced would suggest the presence of a suspended ferric oxide, which would contaminate and color the produced acid red, evidence of iron sulfate. It should be noted that the translators, Schwarz and Kauffman, who were referenced by Karpenko and Norris, also understood that the vitriol involved was iron sulfate.[40]

Paracelsus

Paracelsus (1493-1541) presented many enigmatic descriptions of vitriol. In *The Treasure of Treasures for Alchemists, Concerning the Green Lion,* he referred to vitriol as the "Vitriol of Venus" and the "Green Lion:"

> *Take the Vitriol of Venus, carefully prepared according to the rules of Spagyric Art, and add thereto the elements of water and air which you have reserved. Resolve, and set to purify for a month according to instructions. When the putrefaction is finished, you will behold the sign of the elements. Separate, and you will soon see two colors, namely, white and red. The red is above the white. The red tincture of the Vitriol is so powerful that it reddens all white bodies, and whitens all red ones, which is wonderful.*
>
> *Work upon this tincture by means of a retort, and you will perceive a blackness issue forth. Treat it again by means of the retort, repeating the operation until it comes out whitish. Go on, and do not despair of*

the work. Rectify until you find the true, clear Green Lion [Leo viridis], which you will recognize by its great weight.[41]

In Chapter XII, "General Instruction Concerning the Arcanum of Vitriol and the Red Tincture to be Extracted from It," of *The Aurora of the Philosophers*, Paracelsus spoke of Vitriol in more mystical terms:

Vitriol is a very noble mineral among the rest, and was held always in highest estimation by philosophers, because the Most High God has adorned it with wonderful gifts. They have veiled its arcanum in enigmatic figures like the following: "Thou shalt go to the inner parts of the earth, and by rectification thou shalt find the occult stone, a true medicine." By the earth they understood the Vitriol itself; and by the inner parts of the earth its sweetness and redness, because in the occult part of the Vitriol lies hid a subtle, noble, and most fragrant juice, and pure oil. The method of its production is not to be approached by calcination or distillation. For it must not be deprived on any account of its green color. If it were, it would at the same time lose its arcanum and its power.[42]

The arcanum of vitriol was oil of vitriol, or, in modern terms, sulfuric acid. Paracelsus pointed out that "the earth" spoken of by the philosophers (who the philosophers were Paracelsus did not say. This may have been a general reference to the alchemical tradition) was vitriol. The oil of vitriol was then found within "the earth" after treatment which comprised several steps. In a note to this chapter Paracelsus explained the making of the arcanum of vitriol—sulfuric acid:

... after the aquosity has been removed in coction from vitriol, the spirit is elicited by the application of greater heat. The vitriol then comes over pure in the form of water. This water is combined with the caput mortuum left by the process, and on again separating in a balneum maris, the phlegmatic part passes itself off, and the oil, or arcanum of vitriol, remains in the bottom of the vessel.

(Using the discussion on Basil Valentine above may prove helpful in understanding Paracelsus's description of vitriol and the making of oil of vitriol.)

The *balneum maris*[43] was a distillatory furnace containing water, into which, when hot, a chemical vessel was placed for the putrefaction of the substances contained within the vessel and their consequent separation. The process involved, coction (the heating of a substance at moderate heat for an extended period of time), would, presumably, drive off sulfur dioxide, SO_2 (*Spiritus vitrioli*) and react with water, yielding H_2SO_3, which upon oxidation would form sulfuric acid. The *caput mortuum* may have been *Crocus Martis, the* brownish-red metallic compound ferrous sulfate, $FeSO_4$, which would have been left behind. If a higher heat was used, as perhaps implied by the phrase "... application of greater heat...," then the *caput mortuum* could have been Fe_2O_3 (*colcothar*), since ferrous sulfate decomposes to ferric oxide at temperatures above 480°C. The phlegmatic part would have referred to the structural water of the ferrous sulfate, $FeSO_4 \cdot 7H_2O$.[44]

The purpose of this brief excursion was not to delve into the detailed chemical reaction equations involved, but to translate Paracelsus's language and to point out which vitriol he may have had in mind. In regard to the specific vitriol, this may be confusing since the phrase "Vitriol of Venus" would have implied copper sulfate because copper was associated with Venus. Another possible interpretation might be that the reference was to the vitriol associated with copper mining. In that case, the vitriol would have been iron sulfate, which would be consistent with the other references in Paracelsus's description. The redness, for instance, would imply an oxide of iron, which would have contaminated the preparation of the acid, and the green color would imply the presence of iron sulfate.

In addition to making *Oleum vitrioli*, sulfuric acid, Paracelsus believed that vitriol could transmute iron into copper:

> *A third test is when it transmutes iron into copper. The more perfectly and the more rapidly it does this, the better should it be esteemed in both faculties, for there is great affinity between these two metals... there is a fountain in Hungary, or rather a torrent, which derives its origin from Vitriol, nay, its whole substance is Vitriol, and any iron thrown into it is at once consumed and turned to rust, while this rust is immediately reduced to the best and most permanent copper, by means of fire and bellows.*[45]

What led Paracelsus to this conclusion? Iron is more reactive than copper, and, consequently, it reacts with the copper sulfate solution to displace the copper to form iron sulfate. The chemical reaction is known as a single displacement reaction with the chemical equation of $3CuSO_4 + 2Fe \rightarrow Fe_2(SO_4)_3 + 3Cu$. This is the reaction used in the extraction of copper. Thus, it would have appeared to Paracelsus that the iron was transmuted into copper.

Georgius Agricola (1494-1555)
In his works *De Natura Fossilium* (1546) and *De Re Metallica* (1556), Agricola may have provided the most complete description of the formation and characteristics of *Atramentum sutorium* and the related minerals—chalcitis, misy, sory, and melantria—and the methods for making vitriol. According to *De Natura Fossilium*, *chalcitis* (cuprous pyrite) was the "parent and source" of sory and melantria, which was also called *Atramentum metallicum*. These minerals, in turn, produced *Atramentum sutorium*:[46]

> *Atramentum sutorium is produced both in nature and artificially from water and Sory or Melantria or even Chalcitis. The natural mineral forms from some kind of humor and has congealed like ice either in veins, fractures, or joints of rocks or it has come out from the rock drop by drop, and moving down along channels, has congealed in the form of icicles and hangs from the back of openings or finally it may drop from the back of openings and congeal on the floor. Irrespective of whether it hangs from the back or occurs on the floor, the Greeks call it σταλαχτιχος [Stalactikos] because it has congealed by dropping. There are two artificial varieties, one the Greeks call, correctly πηχτός [pectos], the Latins, concretus. When this variety is either carried from underground workings and poured into rectangular reservoirs or conducted thence along channels, it will congeal due to the cold or heat of the sun. The other variety the Greeks call εφϑός [hephthos], the Latins, coctum or "cooked." The water containing the Atramentum is placed in rectangular basins and then boiled away...*[47]
>
> *Sory and Melantria form from pyrite, which is the source of all the juices; Chalcitis, from Sory; varieties of Atramentum sutorium from Chalcitis, Melantria, and Sory. Some of these minerals occur as white efflorescences, while others are green and even blue. Misy forms as an efflorescence, not only on Sory, Melantria, and Chalcitis but also on all varieties of Atramentum sutorium, both natural and artificial. ... Sory, Melantria, Chalcitis, and Misy are always natural minerals. Sory and Melantria have the same color, gray and black; Chalcitis*

red and copper-colored; Misy, yellowish and golden; Atramentum sutorium, various colors....All five minerals are astringent and acrid, Atramentum sutorium being the most strongly astringent. All have a natural odor similar to that given off by a bolt of lightning, while the odor of Sory is the most penetrating. Atramentum sutorium is soft and tenuous, similar to down or hair; Melantria, similar to plant down and with a certain saltiness. While all five minerals may be light and porous, Sory, Chalcitis, and Misy may occur massive. Sory may be as hard as stone because of excessive congealing, and for that reason, is the most dense; Misy, the most tenuous; Chalcitis, intermediate.[48]

Also, in *De Natura Fossilium*, Agricola gave three varieties of *Atramentum sutorium*—*viride*, *caeruleum*, and *candidum* (green, blue, and white vitriol, respectively). Agricola may have been the first in Europe to mention white vitriol (zinc sulfate), which he described as "A white sort...found, especially at Goslar in the shape of icicles, transparent like crystals."[49] Agricola continued:

> *I will next speak of an acrid, solidified juice which commonly comes from cadmia. It is found at Annaberg in the tunnel driven to the Saint Otto mine; it is hard and white, and so acrid that it kills mice, crickets, and every kind of animal. However, that feathery substance which oozes out from the mountain rocks and the thick substance found hanging in tunnels and caves from which saltpeter is made, while frequently acrid, does not come from cadmia.*[50]

Herbert and Lou Hoover, the translators of *De Re Metallica*, believed Agricola's was the first mention of zinc sulfate in Europe.[51] The translators of *De Natura Fossilium*, Mark and Jean Bandy, concurred, offering Agricola's description of Goslarite (zinc sulfate)[52] as follows:

> *White Atramentum sutorium is called λευκόϊν [leukóin, see also lonchoton above] because it resembles the color of the white violet. It may also be pale to deep green or blue. The finest white variety occurs in the form of icicles at Goslar and resembles transparent quartz. Both the blue and green varieties may be transparent. Because of this transparency, Atramentum sutorium was given the name Vitriolum in olden times.*[53]

In addition to describing the vitriols, Agricola in *De Re Metallica* provided a lengthy discussion of four methods for the preparation of vitriol, in which chalcitis, misy, sory, and melanteria were significant ingredients:

> *Vitriol can be made by four different methods; by two of these methods from water containing vitriol; by one method from a solution of melanteria, sory, and chalcitis; and by another method from earth or stones mixed with vitriol.*[54]

In the recipes for making vitriol, Agricola referenced the use of iron strips in the vats. These strips would have effectively precipitated any copper. Thus, the vitriol produced would have been iron sulfate, which Agricola most likely meant by his use of the term *Atramentum sutorium*.

Agricola discussed alum at length in *De Natura Fossilium*, relating it to, yet distinguishing it from, *Atramentum sutorium*: "...the two minerals can be separated since Alum forms from *Atramentum sutorium*." According to Agricola, alum occurred in both liquid and solid form. One of the liquid genera, the color of milk, smelled of fire, indicating the presence of sulfur. One of the solid varieties, which the Greeks called schisis, formed and grew on mineral veins but also on *Atramentum sutorium*. Pyrites were the parent of both minerals. Agricola followed Pliny in describing the colors of alum as white, grayish-white, or black, and the taste as strongly astringent. The odor of alum was that of fire like "...that given off by stones when struck together..." with the liquid form having the strongest odor. Also, alum was used for dyeing wool–the light-colored alum used for dyeing wool a light color, the dark alum for dyeing wool black. Agricola also pointed out that chalcitis or similar minerals often had to be separated from the liquid alum by heating to produce pure white alum. Here was another parallel between alum and minerals associated with *Atramentum sutorium*.[55]

Martinus Rulandus (1532-1602)

Martinus Rulandus (German physician and alchemist, and follower of the physician/alchemist Paracelsus) summarized the transformations vitriol and *chalcitis* could undergo in his lexicon, in which he introduced another Arabic term, *Zeg* (*Colcothar*), and furthered the understanding of how the substances related to vitriol interacted:

1. *Green Zeg [Arabic]* or *Shoemaker's Black,* or *Vitriol,* or *Chalcanthum (native)* is changed or passes into:
   1. *Misy,* very easily.
   2. After a long time into *Chalcitis,* as regards outward appearance; internally it is still Shoemaker's Black.
   3. *Filaments,* when it is old. Manufactured *Chalcitis* is then wrought from it. Also *Chalcanthum* changed into *Chalcitis* can then be made into *Misy.*
2. *Chalcitis,* or honey-yellow *Zeg,* according to Pliny; brass color, according to Dioscorides (*Zeg* is the name given by the Arabs, who also call it *Colcothar*),[56] has a middle position between *Marchasite* (i.e., *Black Zeg,* or *Pyrites,* or *Black Atrament*), and *Vitriol* (i.e., *Green Zeg,* or *Chalcanthum*), and when old can be changed, and passes into *Sory* very easily. *Sory* and *Melanteria* pass on the other hand into *Chalcanthum Leucoion,* i.e., *White.*[57]

Chalcanthum changed into misy (chalcopyrite—copper iron sulfide) or chalcitis which can further transform into sory (marcasite—iron sulfide).

The lack of differentiation between iron and copper sulfates was further apparent in Rulandus's entry on *Atramentum sutorium.* According to Rulandus, *Atramentum sutorium* was simply labeled vitriol or *chalcanthum.* He did not distinguish between blue or green vitriol in the entry, but he did use terms indicating that he—or rather Paracelsus and his followers—recognized that a difference existed between *Atramentum sutorium* and *Kalkou Anthos,* the true Flower of Copper:

> It is not, however, *Kalkou Anthos,* or true Flower of Copper, as we have before stated. *Chalcanthus,* or *Vitriol,* or *Atramentum sutorium,* is one thing; the ancient Flower of Copper is another, and was obtained, among other ways, from the washings of copper ore, while Flower of Copper has...been given as an alternative name of Verdigris, or Copper-Rust. . . . But with the ancients, *Atramentum* meant Vitriol, that metallic substance which is simply congealed water [acid], having the quality of copper, but differing in its form and nature with the stone to which it adheres.[58]

Further, Rulandus shared Dioscorides's assertion that soft and hard *Atramentum* (it is unclear as to what the difference was between soft and hard

Atramentum) were both *Sutorium*, Shoemaker's Black, and classified according to the type of location in which they were found. Stillatic Vitriol, also called Pinarion and Distallatic, "...concreted from humors which are collected by droppings in mines...." This type was also referred to as the German Distillatic Vitriol. Pecton, concreted and congealed Vitriol, formed in caves and grottoes.

Rulandus also classified Shoemaker's Black or vitriol according to color: white vitriol, green vitriol, and blue vitriol. Referencing Dioscorides, Rulandus affirmed that blue vitriol was the best Stillatic Atrament, describing it further as heavy, close-grained, and translucent.

Finally, in addition to the above treatment of vitriols, Rulandus provided over 30 terms used to reference the vitriols, which, for instance, included such terms as Atramentum Album Tectorium (a hydrate of copper flower), Stalactical Atrament of Goslaria (green vitriol), Copperas of Goslaria (green vitriol), Green hard Stalactical Atrament of Goslaria, and Blue Roman (prepared copperas).[59]

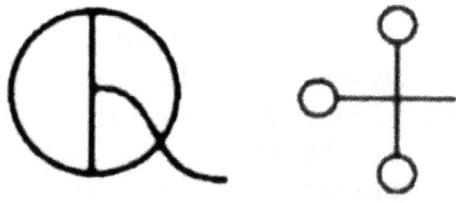

Additional Symbols for Vitriol[60]

Conclusion

Significant to the alchemists and physicians of antiquity and the Middle Ages, the vitriols played an important role in several ways. Physicians used these sulfate compounds as sources of medicines for various treatments from injuries to skin diseases to ulcers. In the commercial world, the vitriols were used in dyeing woolens and leather and in the making of inks. For the alchemists, however, they played a vital role as a source for Oil of Vitriol, sulfuric acid. This acid, along with nitric and hydrochloric acids, provided a means by which the alchemists advanced their knowledge of substances and theories of matter.

As significant as the vitriols were, there was no standard terminology to name and classify them. Thus, the alchemists knew the vitriols by a multitude of names, classified by color (e.g., blue, green, white, and red), by appearance (e.g., Flower of Copper), by method of formation (e.g., artificial vs. natural), by location (e.g., Copperas of Goslaria), or by some combination of these designations (e.g., Green hard Stalactical Atrament of Goslaria). In spite of this variety, the alchemists shared a remarkable consistency in the descriptions of the characteristics and the effects of the vitriols, which abetted the transmission of their observations and discoveries and facilitated advancements in the chemical sciences.

Notes

1. General chemical name encompassing a class of chemical compounds comprising sulfates of certain metals such as iron or copper. From the Latin word *Vitriolus*, meaning "small glass", as the crystals of these sulfates resembled pieces of colored glass.
2. Robert P. Multhauf, *The Origins of Chemistry*, (Langhorne, PA: Gordon and Breach Science Publishers, 1993), 31-32.
3. Marcasite, a term originated in Arabic alchemy, was used as a synonym for pyrites throughout the literature of 16th-century Europe. Vladimir Karpenko and John A. Norris, "Vitriol in the History of Chemistry," *Chem. Listy* 96, (2002), 997-1005.
4. Vladimir Karpenko and John A. Norris.
5. Chalcanthite is widely distributed in arid regions around the world: Cyprus; on Vesuvius, Campania, Italy; Rammelsberg and Goslar, Harz Mountains, Germany; Rio Tinto, Huelva, Spain; in Chile, from Chuquicamata, at Quetena, near Calama, and Copaquire, Antofagasta; in the United States, in Arizona, at the United Verde mine, Yavapai Co.; in the Clifton-Morenci district, Greenlee Co.; the Blue Bird mine, Fierro-Hanover district, Grant Co., New Mexico; the Bluestone mine, Yerington district, Lyon Co., Nevada; Butte, Silver Bow Co., Montana; Ducktown, Polk Co., Tennessee. "Chalcanthite, Handbook of Mineralogy," Mineral Data Publishing, https://rruff.info/doclib/hom/chalcanthite.pdf Retrieved 26 April 2023.
6. A group of triclinic pentahydrated metal (II) sulfates with the general formula, $M2+ SO_4 \cdot 5H_2O$, where M = Mg, Cu, Mn, Fe. Isotypic synthetic selenates are known. Mindat.org, https://www.mindat.org/min-29280.html. Retrieved 27 April 2023.
7. Lesley Brown, *The New Shorter Oxford English Dictionary on Historical Principles*, (Oxford [Eng.]: Clarendon, 1993), 508.
8. Atramentum refers to a black liquid, and sutorium refers to tanning.
9. Vladimir Karpenko and John A. Norris. Also, according to Karpenko, the presence of iron in these substances appears to have remained unrecognized until the sixteenth century.
10. "Handbook of Mineralogy," RRuff Data Base, C:\TYPESET\MDP-WEB\Vol5proc.DVI (rruff.info). Retrieved 07 May 2023.
11. Calcantum—Gk. i.e., "copper flower" probably so-named because of its association with the sulfide ores that were mined for copper—and

from which the mineralogical term chalcanthite was derived. Ancient Greek χάλκανθον (khálkanthon), from χαλκός (khalkós) 'copper', and ἄνθος (ánthos) 'flower, bloom'. See also "Chalcanthite Mineral Data" (webmineral.com), and "Chalcanthite: Mineral information, data, and localities," (mindat.org),

12. From the alteration of the obsolete name chalcosine, from the Greek khalkos, meaning, "copper."
13. Dioscorides, *De Materia Medica*, Tess Anne Osbaldeston, editor/translator, (Johannesburg, South Africa: IBIDIS Press, 2000).
14. Georgius Agricola, *De Re Metallica*, translated from the First Latin Edition of 1556, Herbert Clark Hoover and Lou Henry Hoover, translators, (New York: Dover Publications, Inc., 1950), 572n11-573n11.
15. The Greek name for Cyprus.
16. Misy or copiapite is a hydrated iron sulfate mineral with the formula: $Fe^{2+}Fe^{3+}4(SO_4)_6(OH)_2 \cdot 20(H_2O)$. It is a secondary mineral typically formed by the weathering and oxidation of pyrite in a wide range of rock types, preserved in arid climates; it is rarely associated with fumarolic action. From "The Handbook of Mineralogy," https://rruff.info/doclib/hom/copiapite.pdf. Retrieved 06 May 2023.
17. "Psoricum is made of two parts of calcitis [calcium carbonate] mixed with one part of cadmia [5-84], and pounded into small pieces with vinegar." From *De Materia Medica, Book V*, 116. Cadmia, also called Tutty or Tuttia, was a zinc oxide. See Cadmia- Wikipedia.
18. Dioscorides, *De Materia Medica*, Tess Anne Osbaldeston, editor/translator, (Johannesburg, South Africa: IBIDIS Press, 2000).
19. Ibid.
20. Matteo Martelli, *The Four Books of Pseudo-Democritus,* Sources of Alchemy and Chemistry, *Ambix*, Vol. 60, Supplement 1, (Rutledge: London, 2013), S93-S95, S223n44, S224n45.
21. Pliny the Elder, *Natural History*, Book 34, 32. John Bostock, M.D., F.R.S., H.T. Riley, Esq., B.A. London. Taylor and Francis, Red Lion Court, Fleet Street. 1855. http://www.perseus.tufts.edu/ Retrieved 01 May 2023.
22. Georgius Agricola, *De Re Metallica.*
23. Zinc oxide or the ore bearing zinc oxide.
24. "White iron pyrite", iron sulfide (FeS_2). Marcasite reacts more readily than pyrite under conditions of high humidity. The product of this

disintegration is iron (II) sulfate and sulfuric acid. The hydrous iron sulfate forms a white powder consisting of the mineral melanterite, $FeSO_4 \cdot 7H_2O$.

25. A medicated liquid used for cleaning the mouth.
26. Pliny the Elder, *Natural History, Book 34*, 32. John Bostock, M.D., F.R.S. H.T. Riley, Esq., B.A. London. Taylor and Francis, Red Lion Court, Fleet Street. 1855. http://www.perseus.tufts.edu/ Retrieved 01 May 2023. See also *De Re Metallica*, 573n.
27. Pliny the Elder, Book 35, sections 183-186. http://www.attalus.org/translate/pliny_hn35b.html. Retrieved 07 May 2023.
28. Hugh Chisholm, General Editor, Entry for 'Alum'. 1911 Encyclopedia Britannica. https://www.studylight.org/encyclopedias/eng/bri/a/alum.html. 1910. This information is in the public domain. Retrieved 07 May 2023.
29. Gail Marlow Taylor, *The Alchemy of Al-Razi, A Translation of the "Book of Secrets,"* (North Charleston, South Carolina: Createspace Independent Publishing Platform, 2014), 107.
30. Karpenko and Norris.
31. Taylor.
32. Georgius Agricola, *De Re Metallica*, translated from the First Latin Edition of 1556, Herbert Clark Hoover and Lou Henry Hoover, translators, (New York: Dover Publications, Inc., 1950), 572n11.
33. Albertus Magnus, *Book of Minerals*, translated by Dorothy Wyckoff, Professor of Geology, Bryn Mawr College, 242-244. ALBERTUS MAGNUS *The Book Of Minerals* : Free Download, Borrow, and Streaming : Internet Archive Retrieved 02 May 2023.
34. Ibid, 242n.
35. Pseudo-Geber, "About dissolving liquids and softening oils," *Summa Perfectionis*. See also Vladimir Karpenko and John A. Norris, "Vitriol in the History of Chemistry," *Chem. Listy* 96, 997-1005 (2002).
36. E. J. Holmyard and Richard Russell, *The Works of Geber: Kessinger's Legacy Reprints*, (Kessinger Publishing, 1928), 168.
37. Pseudo-Geber, *Summa Perfectionis, The R.A.M.S. Library of Alchemy, Vol. 9*, (Stuarts Draft, VA: RAMS Publishing Co., 2015), 57, 58.
38. Holmyard and Mandeville, "Avicennae de congelatione et conglutinatione :" E. J. Holmyard and D. C. Mandeville : Free Download, Borrow, and Streaming : Internet Archive. Retrieved 11 May 2023.

39. Chisholm, Hugh, General Editor. Entry for 'Alum'. 1911 Encyclopedia Britannica.https://www.studylight.org/encyclopedias/eng/bri/a/alum.html. 1910. This information is in the public domain. Retrieved 07 May 2023.
40. Karpenko and Norris.
41. Paracelsus, *The Hermetic and Alchemical Writings of Aureolis Philippus Theophrastus Bombast, of Hohenheim, called Paracelsus the Great,* in two volumes, edited by Arthur Edward Waite, (Mansfield Centre, CT: Martino Publishing, 2009), 38, 38n.
42. Ibid, 60, 60n.
43. Martinus Rulandus, *A Lexicon of Alchemy, Containing a Full and Plain Explanation of All Obscure Words, Hermetic Subjects, and Arcane Phrases of Paracelsus*, (Zachariah Palthenus, Bookseller in the Free Republic of Frankfurt, 1612), 69.
44. Karpenko and Norris.
45. Paracelsus, "The Economy of Minerals, Chapter XV, Concerning the Species of Vitriol and the Tests of It," *The Hermetic and Alchemical Writings of Aureolis Philippus Theophrastus Bombast, of Hohenheim, called Paracelsus the Great*, in two volumes, edited by Arthur Edward Waite, (Mansfield Centre, CT: Martino Publishing, 2009), 103.
46. Agricola, *De Natura Fossilium*, translated from the first Latin Edition of 1546 by Mark Chance Bandy and Jean A. Bandy (Mineola, NY: Dover Publications, Inc., 2004), 47-51. See also Agricola's work *De Ortu et Causis Subterraneorum (On the Sources and Causes of What is Underground)*, first published in 1546. In Book I, Agricola stated: "When moisture corrodes cupriferous and friable pyrite, it produces an acid juice from which Atramentum sutorium forms and also liquid alum." In addition, in Book III, he attested that "Not only are Atramentum sutorium and alum made from an acid juice but also sory, chalcitis, and misy. Misy is 'flowers' of Atramentum sutorium, just as Sory is 'flowers' of Melantria."
47. Agricola, *De Natura Fossilium*, 48.
48. Ibid., 49.
49. Ibid. See also Georgius Agricola, *De Re Metallica*, translated from the First Latin Edition of 1556, Herbert Clark Hoover and Lou Henry Hoover, translators (New York: Dover Publications, Inc., 1950), 572n11.

50. Agricola, *De Re Metallica*, 572n11. See also *De Natura Fossilium*.
51. Outside of Europe, however, zinc sulfate was known and referenced much earlier. Ali ben Abbas al-Mujusi (died in Shiraz, Iran, in 994), the author of the medical compendium, *al-Malik*, categorized vitriols as the sulfates of several metals, including copper, iron, and zinc. It was believed that Basil Valentine had also described zinc sulfate as white vitriol. In a later classification found in Oswald Croll's *Basilica Chymica* (first published in Frankfurt, 1610), vitriols were classified by color: green, white, red, and blue, corresponding to iron sulfate, zinc sulfate, and copper sulfate, respectively (copper sulfate's appearance could be either red or blue). Sami K. Hamarneh, "Arabic-Islamic Alchemy-Three Intertwined Stages," *Ambix*, 29:2, 74-87.
52. James Dana (see Dana's *A System of Mineralogy*, 1854) identified the white vitriol described by Agricola as Goslarite, native zinc sulfate. Goslarite, a hydrated zinc sulfate mineral ($ZnSO_4 \cdot 7 H_2O$), was first found in the Rammelsberg mine, Goslar, Harz, Germany. Goslarite belongs to the epsomite group, which also includes epsomite ($MgSO_4 \cdot 7 H_2O$) and morenosite ($NiSO_4 \cdot 7 H_2O$). Goslarite, an unstable mineral at the surface, will dehydrate to other minerals such as bianchite ($ZnSO_4 \cdot 6 H_2O$), boyleite ($ZnSO_4 \cdot 4 H_2O$), and gunningite ($ZnSO_4 \cdot H_2O$). See also Georgius Agricola, *De Re Metallica*, translated from the First Latin Edition of 1556, Herbert Clark Hoover and Lou Henry Hoover, translators, (New York: Dover Publications, Inc., 1950), 572n11.
53. Agricola, *De Natura Fossilium*, 49.
54. Agricola, *De Re Metallica*. The full discussion can be found on pages 572-578.
55. Agricola, *De Natura Fossilium*, 45-47.
56. The usage here does not seem to be the same as that used by Paracelsus, who defined it as fixed vitriol, from which the phlegmatic part has been extracted by distillation until no moisture remains therein. He also associated it with *Caput mortem* and the serpent or green lizard, which devours its own tail. Martinus Rulandus, *A Lexicon of Alchemy, Containing a Full and Plain Explanation of All Obscure Words, Hermetic Subjects, and Arcane Phrases of Paracelsus*, (Zachariah Palthenus, Bookseller in the Free Republic of Frankfurt, 1612), 110.
57. Rulandus, 57.
58. Rulandus, 53-59.

59. Ibid.
60. The alchemical symbols used in this essay were from Philip Wheeler, *Alchemical Symbols, Fourth Edition, R.A.M.S. Library of Alchemy Vol. 21,* (Kansas City, MO:R.A.M.S. Publishing Co., 2018).

Chapter 11

The Mineral Acids

Introduction

The alchemical tradition held that the concept of separation was the path to knowledge.[1] A favored process of separation was that of distillation and solution analysis, in which substances had to be dissolved, commonly by the use of corrosive mineral acids—so-called because they were derived from minerals. These included Spirit of Salt, *Aqua fortis*, Oil of Vitriol, and *Aqua regia*. They were distinguished from organic acids such as citric acid, obtained from citrus fruits, and acetic acid derived from vinegar— which had long been known to the ancient and medieval alchemists—by the lack of carbon. The mineral acids were much stronger and had a greater capacity to separate compounds into their constituent substances or to refine metals from the impurities of their ores.[2] In general, the mineral acids were formed by heating the related mineral, in the presence of other substances, to extreme temperatures in a retort, allowing a reaction of the hot minerals with oxygen and moisture in the air. The resultant acid was then collected as a condensed vapor in the neck of the retort.[3] Tracing the origins and development of these acids, however, has proved to be very difficult. Coded language (*Decknamen*), differences in terminology, uncertain authorship, lack of clarity in recipes, and lack of attribution of the importance of the acid by-products—all obfuscated the early history of these acids.

CHAPTER 11

Spirit of Salt

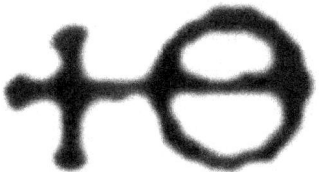

Spirit of salis or Acidum salis

Spirit of Salt, *Spiritus Salis* (Hydrochloric acid, also known as muriatic acid, HCl, an aqueous solution of hydrogen chloride), was one of the first mineral acids to be discovered. It was first produced by the Arabic alchemist al-Razi (full name—Abu Bakr Muhammed ibn Zakariya al-Razi, c. 854/855-925/935, Persian physician, philosopher, and alchemist) who may not have realized he had produced a strong acid. In the early tenth century, he conducted experiments using various recipes found in two sections of his *Secret of Secrets:* "On the sublimation of quicksilver for whitening" and "On the sublimation of quicksilver for reddening." He had used these recipes for the "whitening" of copper or the "reddening" of silver. Most of them included salt and/or sal ammoniac as an ingredient, and three of them also contained vitriol (hydrated sulfates of various metals). In his recipe "On the sublimation of sal ammoniac," he distilled sal ammoniac (ammonium chloride) and vitriol together, producing hydrogen chloride gas.

> *Take equal parts of sweet salt, bitter salt, Ṭabarzad salt, Andarānī salt, Indian salt, salt of Al-Qilī, and salt of Urine. After adding an equal weight of good crystallized sal ammoniac, dissolve it by moisture, and distill the mixture. There will distill over a strong water, which will cleave stone (sakhr) instantly.*[4]

Al-Razi may have stumbled upon a primitive method for producing hydrochloric acid. In most of his experiments, however, he seems to have disregarded the gaseous products, focusing instead on the color changes that could be effected in the residue. In some instances, for example, he focused attention on the redness of the product, which suggested he had calcined the mixture to iron oxide, but ignored the vaporous by-product which would have been hydrogen chloride. According to Robert P. Multhauf, hydrogen chloride was probably produced many times without clear recognition that by dissolving it in water, hydrochloric acid might have been produced. [5]

Drawing on al-Razi's experiments, the *De aluminibus et salibus* ("On Alums and Salts"—an eleventh or twelfth-century Arabic text falsely attributed to al-Razi) described the heating of metals with various salts. In particular, chapter 10 described the heating of mercury, vitriol, and common salt, which resulted in the production of mercury (II) chloride (a corrosive sublimate). In this process, hydrochloric acid actually started to form, but it immediately reacted with the mercury to produce the "corrosive sublimate." [6]

Thirteenth-century Latin alchemists, for whom the De *aluminibus et salibus* was one of the primary reference works,[7] were fascinated by the chlorinating properties of corrosive sublimates, and they soon discovered that when the metals were eliminated from the process of heating vitriols, alums, and salts, strong mineral acids could be directly distilled.

The production of hydrochloric acid as an isolated substance rather than a mixture with nitric acid depended on the use of a more efficient cooling apparatus, which was not developed until subsequent centuries. The need for efficient condensate cooling distinguished the production of hydrochloric acid and nitric acid from that of sulfuric acid. Sulfuric acid did not require the special condensate conditions since its boiling point exceeds that of water. Thus, recipes for the production of hydrochloric acid only appeared in the late sixteenth century, the earliest found in Giovanni Battista Della Porta's (1535–1615) *Magiae naturalis* ("Natural Magic") and in the works of other alchemists like Andreas Libavius (c. 1550–1616), Jean Beguin (1550–1620), and Oswald Croll (c. 1563– 1609).[8]

One of the most detailed recipes from the mid-16th century appeared in John French's *The Art of Distillation, Book 3, Of Minerals*. Again, this later work focused on technique and equipment.

> *Take of the best bay salt as much as you please. Let it be dissolved in spring water and filtered. Mix this with brine in a copper vessel, of the powder of bricks or tiles, twice or thrice as much as the salt before its dissolution was in weight; let the water vapor away over the fire (continually stirring of it) until it is dry. Then put this powder into a glass retort well luted, or an earthen retort, and put it into a furnace (a large receiver joined to it according to art). Then give fire to it by*

degrees until it will bear an open fire, for the space of 12 hours. You shall have a very acid oil or spirit in the receiver. That liquor, being put into a little retort in sand, may be rectified by the vaporizing away of the phlegm. Then keep it for use in a glass very well stopped that no air goes in.[9]

By the early 17th century (c. 1625), Johann Glauber (German-Dutch alchemist and chemist, 1604-1670) had not only developed a recipe for Spirit of Salt, he also became the first person to produce concentrated hydrochloric acid,[10] which he obtained by reacting rock salt with oil of vitriol (sulfuric acid). Glauber achieved several other "firsts," which would also influence the production of the other two mineral acids (see the discussion below), not just the production of hydrochloric acid.

Aqua fortis

Aqua fortis

Aqua Fortis, (nitric acid, HNO_3), also known as *Aqua nitri* or *Aqua valens*,[11] was a powerful oxidizer, capable of oxidizing non-active metals such as copper and silver. Though it is now known that nitric acid was able to dissolve an almost undetectable amount of gold, forming gold (III) ions, the early observations of the alchemists led them to believe that *Aqua fortis* was incapable of dissolving gold. According to them, a much stronger acid was necessary for that.

The earliest known recipe for the preparation of nitric acid was found in *De inventione veritatis* (On the Discovery of Truth), one of the works in the Pseudo-Geber corpus. Unfortunately, dating these writings has been problematic. There may have been several writers involved, but only one has been identified—Paul of Taranto (1290-1330) who has been credited with penning *Summa Perfectionis*.[12] The chapter for his nitric acid recipe was entitled *About dissolving liquids and softening oils*.

Take a pound of Cyprus vitriol [Fe,CuSO₄], a pound and a half of saltpeter, and a quarter of a pound of alum. Submit the whole to distillation, in order to withdraw a liquor which has a high solvent action. The dissolving power of the acid is greatly augmented if it be mixed with some sal ammoniac, for then it will dissolve gold, silver, and sulfur.[13]

The preparation of *Aqua fortis* was also described by Albert the Great (c. 1200-1280) and Ramon Llull (1232-1315/16), who prepared it by distilling a mixture containing niter and green vitriol (Iron (II) Sulfate—$FeSO_4$). He called it "*eau forte.*"[14]

By the sixteenth century, nitric acid was a commonly used substance. Vannoccio Biringuccio (Italian metallurgist and master craftsman of smelting and metalworking, 1480-1539) described in his *Pirotechnia* a method for the purification of nitric acid by adding a small amount of silver, which had the effect of removing the traces of HCl that originated from the KCl sometimes present as an impurity in saltpeter. If this had not been performed beforehand, silver chloride would have formed during the use of the acid, slowing the action and leaving a good deal of silver mixed with the gold residue.[15]

Although the term *Aqua fortis* was already in regular use, Georgius Agricola (1494-1555) referred to it as *Aqua valens* (literally strong water) in his *De re metallica*.[16] This work contained several recipes for this acid, not all of which actually led to nitric acid (some resulted in a mixture of all three strong mineral acids). One of his recipes that did yield nitric acid prescribed the following ingredients: *four librae of vitriol, two and a half librae of saltpeter, half a libra of alum, and one and a half librae of spring water*.[17]

Agricola's and Pseudo-Geber's recipes could yield nitric acid according to the following reactions:

$2CuSO_4 \rightarrow 2\ CuO + 2SO_2 + O_2$
$KNO_3 + SO_2 \rightarrow KO_3SONO$
$2\ KO_3SONO \rightarrow N_2O_3 + K_2SO_4 + SO_3$
N_2O_3 decomposes spontaneously $N_2O_3 \rightarrow NO + NO_2$ or reacts with water
$N_2O_3 + H_2O \rightarrow 2HNO_2$

Then with the disproportionation of HNO_2 produces HNO_3 ($3\ HNO_2 \rightarrow HNO_3 + 2NO + H_2O$)

CHAPTER 11

Later, John French (English physician, 1616-1657) in *The Art of Distillation,* provided this recipe for *Aqua fortis,* which offered details of the technique involved:

> *Take of vitriol calcined two parts and of nitre one part. Grind and mix them well together and put them into a glass retort coated or earthen retort that will endure the fire. Set them into the furnace in an open fire and then, having fitted a large receiver, distill it by degrees the space of 24 hours. Then rectify the water or spirit in sand.*[18]

In 1648 Johann Glauber (German-Dutch alchemist and chemist, 1604-1670) improved the process for the manufacture of nitric acid. He found that by making a slight change in his recipe for hydrochloric acid he could obtain nitric acid. The recipe for Spirit of Salt (hydrochloric acid) called for reacting rock salt with Oil of Vitriol. To make *Aqua fortis,* he reacted nitre (potassium nitrate) and white arsenic (arsenious oxide) with Oil of Vitriol instead—"Plainly after the very same manner as we have taught Spirit of Salt to be prepared, so may also be made 'Aqua fortis'...Instead of salt take niter, and you will have Aqua fortis." The reaction yielded fuming nitric acid, which for many years was referred to as *"spiritus nitre fumans Glauberi."*[19]

Oil of Vitriol

Oil of Vitriol

The history of sulfuric acid (H_2SO_4) was even more difficult to trace, as no reliable, explicit recipe for its preparation was known prior to the sixteenth century. There may be at least two reasons for this. First, sulfuric and nitric acids were often confused (as will be seen below), and the compounds involved are both white and soluble. Second, the production of the acids was often part of a process used to reach other objectives such as the transmutation of metals or the attainment of medicines and elixirs. The alchemists' attention was not as focused on the substances within the processes as much as on the results.[20] Thus, they may not have been aware of the exact nature of the acids discovered. Indeed, there were instances when only descriptive phrases were used such as "... the very solutive water... [extracted with]...the redness of the alembic...[from vitriol, alum, and saltpeter] (from Geber's *De invention)*, or in Geber's *Liber fornacum:* "...our dissolving water...." Abulcasis (Arab Andalusian physician, surgeon, and chemist, 936-1013) in a recipe used the term "an acute water." Or Peter of Spain (13th century), in *Thesaurus pauperum,* produced from sal ammoniac, vitriol, yellow arsenic, and verdigris [copper acetate], a distillate that "... penetrates bones and all metals..." and ". . . cauterizes like fire. . . ."[22]

Sulfuric acid was perhaps first observed in the Muslim world of the 9th and 10th centuries, when alchemists produced it from crystals of sulfate salts—particularly copper sulfate or iron sulfate, also known as the Green Lion (*leo viridis*)[23]—materials known as vitriols.[24] Thus sulfuric acid was called Oil of Vitriol, Liquor of Vitriol, or Spirit of Vitriol. Al-Razi was credited with the discovery of sulfuric acid. According to his work, *Kitab Sirr Al-Asrar* (Book of Secret of Secrets), he obtained sulfuric acid through the use of iron and copper sulfates under distillation.

Though the works of Arabic alchemists had been translated into Latin in Moorish Spain, the knowledge of sulfuric acid apparently was not disseminated to the rest of Europe until the 13th century. Chapter XIII "Of Medicines citrinating [yellowing] Luna," found in the second part of Pseudo-Geber's *Summa perfectionis,* is considered to be the earliest known recipe for sulfuric acid in Europe.[25]

> *Luna is likewise citrinated with a Solution of Mars: but the necessity of this Work induces us first to calcine it, and then to fix, which is an abundance of labour. Afterward, we administer it with the same Preparation, and the same Projection, pouring it upon the Substance of Luna. And yet thence results not a splendid bright Colour, but dull and livid, with a mortiferous Citrinity. But the way of coloring it, which is made by Vitriol, or Copperas, is thus: ℞ of either of them a certain Quantity, and sublime as much thereof, as can be sublimed, until with great expression of Fire it be sublimed. After this again sublime this Sublimate with Fire appropriate to it, that of it part after part may be fixed, until its greater part is fixed. But afterward, it*

must be warily calcined, that a greater Fire may be administered for its Perfection. This being done, it is dissolved into a most Red Water, that hath not its peer.[26]

(There has been some doubt as to whether this recipe would produce sulfuric acid. According to Vladimir Karpenko (Prof. in the Department of Physical and Macromolecular Chemistry at Charles University, Prague), William Newman, in his work on Pseudo-Geber, stated that this was not a recipe for sulfuric acid since copper sulfate decomposes at 700°C to cupric oxide; further heating to 1050°C would produce cuprous oxide, a red compound often used as a pigment. Karpenko, however, referring to the works of Soukup and Mayer, and Schröder, maintained that it was possible Geber had produced sulfuric acid.)[27]

There were also vague allusions to sulfuric acid in the work of Vincent of Beauvais (d. 1264) and in the *Compositum de compositis* ascribed to Albertus Magnus (c. 1200-1280)[28] found below. In both cases, the description concerns the distillation of alum (aluminum sulfate).[29]

Take two pounds of Roman vitriol, two pounds of saltpeter, and one pound of calcined alum. Grind well, mix perfectly, put them into a glass alembic, distill the water by the regulations, seal tight the joints, for fear that the spirits might vanish. Begin by a gentle fire, then increase the heat; then heat by tinder until the apparatus turns white in order that all spirits distill. Then cease the fire, let the furnace cool down; place this water carefully aside; it is the solvent of the Moon; conserve it for the Opus; it dissolves silver and separates from gold. It calcines Mercury and the Crocus of Mars; it imparts to the skin of man a brown coloration, of difficult removal. It is the primal water of the philosophers, perfect to the first degree. You will prepare three pounds of this water.[30]

This recipe, similar to those of Pseudo-Geber and Agricola, probably yielded nitric acid. It was possible, however, that this recipe produced a mixture of acids. Sulfuric acid, for example, could be obtained according to the following: Roman vitriol (also known as blue vitriol)—cupric sulfate ($CuSO_4 \cdot 5H_2O$)—would release all of its water once the temperature reached 250°C, and $CuSO_4$ would break down according to the reaction equation $2CuSO_4 \rightarrow 2\,CuO + 2SO_2 + O_2$. SO_2 then would react with the water (H_2SO_3), and by slow oxidation in air yield sulfuric acid: $2H_2SO_3 + O_2 \rightarrow 2\ H_2SO_4$. The presence of saltpeter (potassium nitrate—KNO_3) and vitriol together, however, was also a common method for producing nitric acid (see the discussion of Agricola above).

Andreas Libavius (c.1556-1616) wrote at length about Spirit of Vitriol (Vitriolgeist) in his book *Alchemia* (1597), in which he described both white and red varieties. In his opinion, the red spirit was pure Oil of Colcothar—Fe_2O_3 precipitated during the reaction—or a red liquor remaining after the separation of a white spirit.[31] The red color suggests that the acid was sulfuric acid contaminated with ferric oxide, which would point to the use of iron sulfate in the production of sulfuric acid. The ferric sulfate would decompose above 480°C ($Fe_2(SO_4)_3 \rightarrow Fe_2O_3 + 3\ SO_3$), leaving behind the *caput mortuum*—colcothar (Fe_2O_3).

Caput mortuum

A recipe (dated around 1602) attributed to Basil Valentine—most likely written by Johann Thölde (?-1624)[32]—would support Libavius's conclusion.

> *If you get such a deeply graduated and well-prepared mineral, called Vitriol [$FeSO_4$], ... put it into a well-coated retort, drive it gently at first, then increase the fire, there comes in the form of a white spirit of vitriol [SO_3] in the manner of a horrid fume, or wind, and cometh into the receiver as long as it hath any material in it ... if you separate and free this expelled spirit well and purely per modum distillations, from its earthy humidity [H_2O], then in the bottom of the glass you will find the treasure, and fundamentals of all the Philosophers, and yet known to few, which is a red Oil, as ponderous in weight, as ever any Lead, or Gold may be, as thick as blood, of a burnt fiery quality.*

The "red Oil" suggests a viscous liquid, which may have resulted from the suspension of red ferric oxide.[33] Similarly, Paracelsus described the distillation of colcothar that had already been used for the production of Spiritus Vitriol and the blood-red, oily liquid (*Oleum vitrioli*).

Later writings began to offer more details regarding the techniques used in producing these acids. For instance, John French (1616-1657, English physician)

CHAPTER 11

recommended the following recipe for the production of Oil of Vitriol, which provided insight into both the techniques and the equipment involved:

> *Take of Hungarian,*[34] *or the best English vitriol, as much as you please. Let it be melted in an earthen vessel glazed, with a soft fire, that all the moisture may exhale, continually stirring it, until it be brought into a yellow powder which must be put into a glass retort well luted or an earthen retort that will endure the fire. Fit a large receiver to the retort and close the joints well together. Then give it fire by degrees until the second day. Then make the strongest heat you can until the receiver which before was dark with fumes is clear again. Let the liquor that is distilled off be put into a little retort, and the phlegm be drawn off in sand. So will the oil be rectified which is most strong and ponderous, and must be kept itself. Many call that phlegm which is drawn off in rectifying, the spirit of vitriol.*[35]

In addition, and perhaps prior to his improvement on the production of nitric acid, Glauber achieved a simpler, more efficient production of sulfuric acid by burning sulfur with potassium nitrate (KNO_3).[36]

Though sulfuric acid—the product of destructive distillation of vitriol or alum—was easy enough to produce, obtaining an apparatus capable of withstanding such a corrosive at high temperatures was difficult. The early alchemists–strangely enough–seemed to have had access to such an apparatus while Robert Boyle (1627-1691), many years later, with improved glass containers, often complained about the inability of his apparatus to withstand the severe conditions.[37]

Aqua Regia

Aqua Regia

An important result from the discovery of the mineral acids was *Aqua Regia* ("Royal water" or "Regal water"),[38] a mixture of nitric acid and hydrochloric acid, capable of dissolving gold, something neither constituent alone could do.[39] Well-known by metallurgists throughout the Middle Ages,[40] "Regal Water" was appropriately named by alchemists because it dissolved the noble metals gold and platinum.[41]

Gold   Platinum

In European alchemy, *Aqua regia* was first described in Pseudo-Geber's *De inventione veritatis* ("On the Discovery of Truth", after c. 1300), in which the corrosive mixture was prepared by adding ammonium chloride to nitric acid.[42] In addition to the reference in Chap. XXIII About dissolving liquids and softening oils (see above discussion on *Aqua fortis*), the addition of sal ammoniac to nitric acid is also mentioned in the following:

> *Put the Powder upon a Porphyry Stone, and add to it two parts of Ammoniac prepared, and one part of Mercury sublimed, grind all very well together, and imbibe the Mixture with Water of Salt-Alkali, or the Water of Salt-Peter (if you find not Zoza, or Zoda)*[43] *and when imbibed, put it to be distilled with gentle Fire, by which Extract the whole Water, so that what remains in the Bottom may be as melted Pitch: then revert the same Water upon it; and this do thrice, repeating the same labor. (Chap. XVII, "Of White Medicines for Jupiter, and Saturn")*[44]
>
> *The first Solar Medicine is thus made. Calcine Sol, amalgamating first with Mercury, and as in Luna express the Mercury through a Cloth: then grind it with twice so much as itself is of Common-Salt prepared, and set the whole over a gentle Fire, that the remaining Mercury may recede. Extract the Salt with sweet Water, and dry the Calx, from which sublime as much of Sal ammoniac, reverting the*

CHAPTER 11

sublimed Salt upon it four times; then dissolve it in the Water of Vitriol, and Peter [a reference to nitric acid], and Jamenous Allom, as is taught in the end of this Book. (Chap. XVIII, "Of Solar Medicines for Jupiter and Saturn")[45]

Because of its ability to dissolve gold, *Aqua regia* was instrumental proof that the alchemist was on the right path toward producing the Philosopher's Stone. To "make the fixed volatile and the volatile fixed" was a guiding axiom for alchemists searching for the Philosopher's Stone. Few substances were more fixed or nonvolatile than gold, so dissolving it would be a significant step toward achieving the axiom of the "ancient sages," a definite sign that one was on the right path.[46]

Basil Valentine (15th-century alchemist) in his *Twelve Keys,* certainly believed he was on the right path. In the *Second* and *Third Keys* within this work, he described in great detail (albeit in code) the production and use of *Aqua regia* to dissolve gold. The second key provided his recipe for *Aqua regia*.

Emblem from *The Twelve Keys of Basilius Valentinus*

... the bath in which the bridegroom is placed must consist of two hostile kinds of matter that purge and rectify each other by means of a continued struggle. For it is not good for the eagle to build her nest on the summit of the Alps because her young ones are thus in great danger of being frozen to death by the intense cold that prevails there. But if you add to the eagle the icy dragon that has long had its habitation upon the rocks and has crawled forth from the caverns of the earth and place both over the fire, it will elicit from the icy dragon a fiery spirit, which, by means of its great heat, will consume the wings of the eagle and prepare a perspiring bath of so extraordinary a degree of heat that the snow will melt upon the summit of the mountains and become a water with which the invigorating mineral bath may be prepared ... (from *The Second Key*).[47]

The *Third Key* described how *aqua regia* acted on gold.

Emblem from *The Twelve Keys of Basilius Valentinus*

CHAPTER 11

> *... our fiery sulphur must be overcome by means of our prepared water. But, after the water has vanished, the fiery life of our sulphurous vapor must triumph, and again obtain the victory. But no such triumph can take place unless the King imparts great strength and potency to his water and tinges it with his own color, that thereby he may be consumed and become invisible, and then again recover his visible form ...* (from *The Third Key*).[48]

Fortunately, Lawrence Principe has decoded the numerous Decknamen in these passages. The "...bath in which the bridegroom is placed..." was prepared by "...two hostile kinds of matter, also referred to as the Eagle and the Dragon. Valentine had previously identified the Eagle as sal ammoniac, and by Principe's detective work, the Dragon has been identified as potassium nitrate. The Eagle was an appropriate alias for sal ammoniac since the Eagle flies through the air and sal ammoniac sublimes (volatizes) easily upon mild heating. ("Volatize" derives from the Latin "volare," i.e., to fly). The iciness of the Dragon further supported the notion that it was saltpeter since potassium nitrate lowers the temperature of water as it dissolves. The "... fiery spirit..." that was driven out of the Dragon by heat was a reference to nitric acid, further proof that the Dragon was potassium nitrate. Thus, when ammonium chloride and potassium nitrate were mixed ("add to the Eagle the icy Dragon"), placed in a retort in a furnace ("place both over the fire"), and subjected to an intense heat ("great heat"), a strong reaction took place ("two hostile kinds of matter that purge and rectify each other by means of a continued struggle"), and a highly corrosive acid distills ("the invigorating mineral bath")—*Aqua regia*. (How the addition of sal ammoniac to nitric acid produces the combination of hydrochloric acid and nitric acid was discussed above).[49]

The directions in the *Third Key* described the action of the prepared acid ("water") on the purified gold ("Sulfur"). The gold was dissolved by the acid into a transparent solution ("consumed and become invisible"). The gold, then, reappears ("recover his visible form"), which suggested that the solution should be evaporated, leaving behind a residue of gold chloride. (Gold chloride is unstable in the presence of heat, so when its solution evaporated, the residue decomposed quickly to produce gold once again—thus the King's visible form reappeared.)[50]

Later, alchemists moved away from the use of Decknamen to provide clearer descriptions, and in some cases, quantification of ingredients. Michael

Sendivogius's (1566 – 1636, Polish philosopher, alchemist, and physician) recipe, for instance, made explicit reference to the substances used: fixed salt (potassium carbonate), volatile salt (ammonium chloride), and Spirit of Niter (nitric acid). The result was a mixture of nitric and hydrochloric acids.[51]

John French went further and provided the following instructions for making *Aqua regia*, specifying the necessary quantities of the ingredients:

> *Take of nitre two parts, salt armoniac one part, and the powder of flints three parts. Put them into a glass retort coated or earthen retort that will endure the fire. Distill them by degrees over a naked fire for the space of 24 hours. Take it out and rectify it. This water will dissolve gold.*

Or

> *Take of spirit of nitre as much as you please. Put a dram of crude nitre to every ounce of it and it will be as strong as any aqua regia. This water will dissolve gold.*[52]

The knowledge of mineral acids, such as hydrochloric acid, would be of key importance to seventeenth-century chemists like Daniel Sennert (1572–1637) and Robert Boyle (1627–1691), who used the mineral acids' capability to rapidly dissolve metals in their demonstrations of the composite nature of bodies.[53]

Conclusion

Tracing the origins and early history of the mineral acids was quite problematic—especially for sulfuric acid. Questions of authorship, use of coded language, application of the same terms having different meanings or different words having the same meaning—all contributed to the difficulties. In addition, the recipes were often vague, providing instructions which could lead to the production of any one or a number of the mineral acids. Moreover, the alchemists, themselves, were often uncertain—may even have been unaware, as in the case of hydrochloric acid—that they had produced a particular acid at all. They were focused more on other objectives and did not give adequate consideration to the by-products of their experiments, which may have been mineral acids.

The alchemists of antiquity and the Middle Ages, then, were up against monumental obstacles. In their pursuit to understand nature, they had to not only create new tools to uncover nature's secrets, but they had to invent a brand-new language to convey what they were learning to their contemporaries. Following their philosophical perspective that knowledge was gained through separation, they sought the means by which nature could be dissected, separated—just as creation, itself, was an act of separation. The gradual discovery of the mineral acids, which came by hit or miss, trial and error, helped them in their endeavor. Eventually, these acids became the toolkit that facilitated the separation of metals from ores, metals from metals, and ultimately made possible the breakdown of compounds into their components.

Notes

1. Keith Schuette, "The Elusive Alkahest," *Hexagon*, Fall 2022, Vol. 113, No. 3.42-45.
2. Robert P. Multhauf, *The Origins of Chemistry*, (Langhorne, PA: Gordon and Breach Science Publishers, 1993), 140-141. The mineral acids were made clearly only about three centuries after al-Razi, in the works of the European alchemists.
3. Cathy Cobb, Monty L. Fetterolf, and Harold Goldwhite, *The Chemistry of Alchemy: From Dragon's Blood to Donkey Dung—How Chemistry was Forged* (Amherst, NY: Prometheus Books, 2014), 115–117.
4. Gail Marlow Taylor, *The Alchemy of Al-Razi: A Translation of the "Book of Secrets"* (CreateSpace Independent Publishing Platform, 2015), 139–140.
5. Multhauf, *The Origins of Chemistry* (Langhorne, PA: Gordon and Breach Science Publishers, 1993), 141–142, 142n.
6. Multhauf, 161–162. The *De aluminibus et salibus* is believed to have been translated in Spain by Gerard of Cremona (1114–87) and became the most influential source of information to the early Latin alchemists. *De aluminibus et salibus* (ch 81): sal ammoniac '...dissolves mercury and changes it into a running water when they are exhaulted together.' (ch 11) This exhaultation results from the distillation of a mixture of mercury, sal ammoniac, and alum. (ch 10) The same product was obtained by heating mercury, vitriol, and common salt. The product in either case would have been the chloride of mercury, corrosive sublimate ($HgCl_2$), which results from the incipient generation of hydrochloric acid and its immediate reaction with mercury, the reaction being terminated by the sublimation of the corrosive sublimate.
7. Multhauf, 160–162.
8. Multhauf, 204, 208n29.
9. John French, *The Art of Distillation: The R.A.M.S. Library of Alchemy, Vol. 15* (Stuarts Draft, VA: R.A.M.S. Publishing Co., 2016), 128.
10. Claudia Flavell-While, Johann Glauber-Alchemy to Modern Chemistry, *The Chemical Engineer*. https://www.thechemicalengineer.com/features/cewctw-johann-glauber-alchemy-to-modern-chemistry/. Retrieved 23 Feb 2023.
11. Martinus Rulandus, *A Lexicon of Alchemy, Containing a Full and Plain Explanation of All Obscure Words, Hermetic Subjects, and Arcane*

Phrases of Paracelsus, (Zachariah Palthenus, Bookseller in the Free Republic of Frankfurt, 1612), 35.
12. Vladimir Karpenko and John A. Norris, "Vitriol in the History of Chemistry," *Chem. Listy* 96, 997-1005 (2002).
13. The addition of sal ammoniac to the distillate leads to *aqua regia* (a mixture of HNO3 + HCl, in a proportion of 1:3).
14. "The history of chemistry:" Thomson, Thomas, 1773-1852 : Free Download, Borrow, and Streaming : Internet Archive. Thomas Thomson (chemist), The History of Chemistry (1830) Vol. 1, p. 40. Retrieved 11 Feb 2023. See also Chisholm, Hugh, ed. (1911). "Nitric Acid." Encyclopedia Britannica. Vol. 19 (11th ed.). Cambridge University Press. 711–712.
15. Vannoccio Biringuccio, *Pirotoechnia,* edited by Cyril Stanley Smith and Martha Teach Gnudi, (Cambridge, Mass.: The M.I.T. Press, March 1966). 186. Pirotoechnia was considered the earliest printed work to cover the whole field of metallurgy. (ix—xi). Biringuccio also used Aqua de Partire (parting acid) as another term for Aqua fortis and often used Aqua acuta (sharp water), which was a general term for acids (see note page 183).
16. Georgius Agricola, *De Re Metallica*, translated by Herbert Clark Hoover and Lou Henry Hoover (New York: Dover Publications, 1950), 439. Originally published posthumously in 1556, this was the first book on mining to be based on field research and observation. The book was translated into English in 1912 by former US President Herbert Clark Hoover and his wife, Lou Henry Hoover. Herbert Hoover, a graduate of Stanford University, was himself a mining engineer.
17. Ibid., 439.
18. French, 137.
19. Claudia Flavell-While, "Johann Glauber—Alchemy to Modern Chemistry," The Chemical Engineer. https://www.thechemicalengineer.com/features/cewctw-johann-glauber-alchemy-to-modern-chemistry/ Retrieved 23 Feb 2023.
20. Multhauf, 226–228.
21. The Latinized version of Abū al-Qāsim Khalaf ibn al-'Abbās al-Zahrāwī al-Ansari (c. 936–1013), Arab Andalusian physician, surgeon, and chemist. Retrieved 14 Feb 2023 from https://en.wikipedia.org/wiki/Al-Zahrawi.

22. Multhauf, 172–173, 207n.
23. Multhauf, 195. See also the tract *"De leone viride"* in the *Sanioris medicinae*, which opens with a recipe for the generation of sulfuric acid.
24. From the Latin *vitreus* ("crystal").
25. Karpenko and Norris. See also Robert P. Multhauf, *The Origins of Chemistry*, Langhorne, PA: Gordon and Breach Science Publishers, 1993), 173.
26. E. J. Holmyard and Richard Russell, *The Works of Geber: Kessinger's Legacy Reprints*, (Kessinger Publishing, 1928), 168.
27. For a complete discussion of this issue and the reaction equations involved, see Vladimir Karpenko and John A. Norris, "Vitriol in the History of Chemistry," *Chem. Listy* 96, 997-1005 (2002).
28. If this recipe was truly an instruction for the making of sulfuric acid, it would predate the recipe found in the *Summa Perfectionis* of Pseudo-Geber.
29. The alchemists of the Middle Ages did not always discriminate between aluminum sulfate and iron sulfate.
30. Albertus Magnus, *The Compound of Compounds*, translated from the French by Luc Villeneuve. Introduction by Adam McLean, Hermetic Research Series, (Glasgow, 2003).
31. Karpenko and Norris.
32. Johann Thölde was an owner of a salt-works in Thuringia, a state in central Germany.
33. Karpenko and Norris.
34. Hungarian vitriol—probably iron sulfate.
35. French, 131-132.
36. CRGsoft.com, "Sulfuric Acid: What It Is, Properties, Uses, and Characteristics," https://crgsoft.com/sulfuric-acid-what-it-is-properties-uses-and-characteristics/ Retrieved 03 Feb 2023.
37. Multhauf, 204, 204n.
38. *Aqua regia*, a fuming liquid, is colorless, but it turns yellow, orange, or red within seconds from the formation of nitrosyl chloride and nitrogen dioxide.
39. Nitric acid will dissolve an almost undetectable amount of gold, forming gold(III) ions (Au3+). The hydrochloric acid chloride ions (Cl−), which react with the gold ions to produce tetrachloroaurate (III) anions ($[AuCl_4]-$), also in solution. The reaction with hydrochloric acid

is an equilibrium reaction that favors the formation of tetrachloroaurate (III) anions, which results in the removal of gold ions from solution and allows further oxidation of gold to take place. (See "Gold, Gold Alloys, and Gold Compounds," *Ullmann's Encyclopedia of Industrial Chemistry*)

40. Multhauf, 223.
41. M. E. Weeks, "Discovery of the Elements," *Journal of Chemical Education*, 1968, 385–407. The first European reference to platinum appears in 1557 in the writings of the Italian humanist Julius Caesar Scaliger as a description of an unknown noble metal found between Darién (a district of Panama) and Mexico, "which no fire nor any Spanish artifice has yet been able to liquefy".
42. Karpenko and Norris.
43. Salt-Alkali made of Zoda.
44. E. J. Holmyard and Richard Russell, *The Works of Geber,* Kessinger's Legacy Reprints (Kessinger Publishing, 1928), 218.
45. Holmyard, 219.
46. Lawrence M. Principe, *The Secrets of Alchemy,* (Chicago: University of Chicago Press, 2013), 152.
47. Basilius Valentinus, *The Twelve Keys*: *The R.A.M.S. Library of Alchemy, Vol. 1 (*Stuarts Draft, VA: The R.A.M.S. Publishing Co., 2015), 41-44.
48. Valentinus, 45-46.
49. Principe, 147-151.
50. Ibid.
51. Cathy Cobb, Monty L. Fetterolf, and Harold Goldwhite.
52. French, 137.
53. William R. Newman, *Atoms and Alchemy: Chymistry and the Experimental Origins of the Scientific Revolution* (Chicago: University of Chicago Press, 2006), 98.

Chapter 12

The Elusive Alkahest

Originally Published in the *Hexagon*, Fall 2022, Vol. 113, No. 3

In the beginning, God created the heavens and the earth. The earth was without form and void, and darkness was upon the face of the deep; and the Spirit of God was hovering over the face of the waters. Then God said, "Let there be light;" and there was light. And God saw the light, that it was good; and God divided the light from the darkness. (Gen. 1:1-4 KJV)

To Medieval and Renaissance alchemists, the opening words of the Bible were evidence that separation was the beginning of knowledge. According to alchemists, especially Paracelsus, fundamental knowledge was the knowledge of separation. Thus, the processes of separation—distillation, calcination, and sublimation—were crucial to understanding nature. By means of these processes, the elements could be separated from their "mixts," fifth essences or quintessence,[1] could be freed, and healing and perfecting secrets could be discovered. ([2]) The application of this epistemology led to an increased understanding of medical alchemy—also known as iatrochemistry—transmutational alchemy, and, through the pursuit of the alkahest, solvents. To discover a substance that could separate all other substances into their prime components would be key to this epistemology.

CHAPTER 12

A Key to the Search for Chemical Knowledge
The distillation and sublimation processes were favored among alchemists to use in their experiments designed to advance their understanding of matter and chemical knowledge. These processes were dependent on the use of solvents. *Aqua fortis* (nitric acid), for example, was used to dissolve most metals, but would not dissolve gold. Samuel Dulcos (founder of the Parisian Royal Academy of Sciences' chemical program and laboratory in 1666), drawing upon Paracelsian and Helmontian insights, believed solution analysis was the ultimate tool for discerning the nature of substances. He held that the application of a universal solvent was the road to transcending the products of distillation—the Paracelsian Tria Prima, the Aristotelian four elements, or combinations thereof—and achieving an authentic chemical resolution of mixts into their true principles or elements.[3] Ideally, this universal solvent would be capable of dissolving any other substance, including gold, without altering or destroying its fundamental components.[4] To have such a solvent was therefore desirable, encouraging a search for the universal solvent—the alkahest. To some alchemists and chymists, the endgame was to recombine the primal elements freed by the alkahest into the Philosopher's Stone or an elixir, which could transmute other substances; while for others, the goal was to recover primary constituents for the preparation of medicines.

Etymology and Nomenclature of the Alkahest
Because the pursuit of the universal solvent was undertaken by many alchemists, the alkahest became known and understood by various names and concepts. The term "alkahest" was first coined by Paracelsus, and according to George Starkey (American alchemist, 1628-1665), was derived from the German *al-gehest*, i.e., all spirit,[5] which is conjecture since Paracelsus left no trace of the origins of the word. Other origins suggested included: *alhali est,* the German word *al gar heis,* or *Al zu hees,* meaning "very hot"—all suggested by Johann Rudolph Glauber.[6]

Beyond its etymology, the alkahest was also subject to disparate names, purposes, and natures. Paracelsus believed it was the elusive Philosopher's Stone.[7] He also used the word to reference a particular medicine for the liver.[8] Herman Boerhaave (Dutch botanist and physician—1668-1738), on the other hand, in his textbook *Elementa Chymiae* (1732), did not believe it was the Philosopher's Stone but was of greater importance and value, since it would supposedly procure both riches and 'the most efficacious remedies' for bodily ills. Henry Oldenburg (one of the creators of the modern scientific peer review) made experimental connections between the alkahest and the liquid in the lymphatic vessels of animals that was discovered by Tobias Ludwig Kohlhans (physician, 1624-1705).[9]

Joan Baptista van Helmont (Belgian physician, 1579-1644) referred to the alkahest as an incorruptible dissolving liquid that could reduce any substance (a material called sal circulatum, "circulated salt," by Paracelsus), including metals and minerals, into its elementary constituents.[10] Other names Helmont assigned to the alkahest included arcanum of fire,[11] "immortal," "maccabean fire,"[12] ignis gehenna (since it reduced bodies that resisted 'Vulgar Fires' into their primary constituents), and primum Ens Salum (salt exalted to its highest degree). Further, van Helmont saw the implications of the alkahest for remedies of bodily ills. Reflecting on van Helmont's claims regarding the alkahest, Robert Boyle (natural philosopher, 1627-1691)[13], in *The Sceptical Chymist* (1661), stated...

> *To this liquor he ascribes, (and that in great part upon his own experience) such wonders, that if we suppose them all true, I am so much the more a friend to knowledge than to wealth, that I should think the alkahest a nobler and more desirable secret than the philosopher's stone itself.*[14]

George Starkey, in *The Secret of the Immortal Liquor Called Alkahest or Ignis-Aqua*—written under the pseudonym Eirenaeus Philalethes—described the alkahest as a "... Catholic and Universal Menstruum ...," a "Fiery Water" (Ignis Aqua), and "... an uncompounded and immortal essence, which is penetrative, resolving all things into their first Liquid Matter. ..." In regard to its substance, Starkey considered this Immortal Liquor a "noble circulated salt,"—not any ordinary salt made liquid by a simple solution, but "... a saline spirit which heat cannot coagulate by evaporation of the moisture, but is of a spiritual uniform substance, volatile with a gentle heat, leaving nothing behind it...." Further, he attested that nothing can resist its dissolving power while itself is not changed, remaining whole in its nature and form.[15]

The Mechanism of the Alkahest

The concept of the alkahest originated in the Paracelsian corpus, but received its mature form primarily through the work of Joan van Helmont and his successors. Rejecting the Paracelsian Tria Prima theory of material, which held that Mercury, Sulfur, and Salt were the basic elements of all substances (in contradistinction to Geber's theory that all matter was composed of just Mercury and Sulfur), van Helmont favored monism, which purported that water was the basic substratum of all substances, and the Tria Prima were just proximate ingredients. According to van Helmont, fire destroys substances by converting them to gas (from the Greek word for chaos, meaning a non-condensable substance more subtle than vapor), which rises to the upper layer of the atmosphere, where exposed to extreme cold, condenses and returns as elemental water that falls as rain. The alkahest performs this return to water more quickly and efficiently. Using heat and the alkahest, a substance is first decomposed into its proximate ingredients (Tria Prima), and then upon further heating is reduced to water. The crucial point of the process was to stop at just the right point and distill off the alkahest so the "first essence" (ens primum) of the dissolved substance would be left behind as a crystallized salt. This ens, then, contained the pure substance, free from impurities or noxious properties. For medical alchemy, this meant the resultant salt contained concentrated medicinal powers.[16]

Along with the Helmontian theory, corpuscularianism was used to describe the specifics of the mechanism of the alkahest. According to Helmont and Robert Boyle, the alkahest was composed of extremely small, homogeneous corpuscles which moved between the corpuscles of other materials and mechanically separated them without altering the base materials or itself. The minuscule particle size of the alkahest allowed it to penetrate into the pores of even the densest substances, such as gold, and sever the union of the substance's corpuscles. Divided into such minute particles, the corpuscles could not support the work of the "semina"[17] that enabled the substance to have unique qualitative characteristics. The continuation of this division and prohibition of the "semina's" action would eventually reduce the substance into primordial water. Also, van Helmont asserted that the alkahest's corpuscles were uniformly very small, which allowed the alkahest to be distilled off without leaving any residue. This made the alkahest reusable and distinct from ordinary corrosives, which are altered by the substances they act upon. If a substance continued to be divided into increasingly smaller parts, van Helmont warned that the substance would be reduced to elementary water. He proposed, however, that partial subtilization (division into smaller particles) of a substance could change some of its properties, but the substance would remain essentially the same.[18]

Van Helmont viewed the alkahest as more than merely a chemical used in the manipulation of matter, however. He did not relate the alkahest to the Philosopher's Stone or the transmutation of metals into gold. The alkahest was, in the alchemical mind, more than a "solvent" in the modern sense—having the power to dissolve, resulting in a solution. This "ponderous liquid" was an important means for the preparation of medicines and for revealing secrets hidden in natural bodies. Beginning with Paracelsus and maturing through van Helmont, this understanding of the alkahest gained ascendancy in iatrochemistry over the transmutational view. The alkahest was seen as the means to recover primary elements from a body, including the seminal virtues that provided therapeutic efficacy. With the alkahest, the physician would be able to cure incurable diseases and prepare a medicine for prolonging human life. For instance, it was believed that by dissolving Ludus, a mineral, with the alkahest, a remedy could be found for kidney stones, a condition at that time believed to be incurable.[19]

Later, Robert Boyle put a mechanical twist to van Helmont's theory.[20] In agreement with van Helmont, Boyle argued that the alkahest works by dissecting substances into such small particles that they become liquid and even volatile. This dovetailed into Boyle's theory that volatility was related to small particle size. He asserted that small particle size was amenable to agitation and motion, which rendered the substance fluid. Boyle's mechanical approach attempted to answer a critique of the Helmontian theory. If the alkahest could be removed intact from the substances it was acting on, this meant that the corpuscles of the alkahest did not combine with those of the dissolved substance. Boyle argued, then, that the characteristics of volatility and fluidity induced into the dissolved substance could not be attributed to any amount of alkahest that remained. Thus, he concluded, the resultant volatility and fluidity of the substance were due to a change of state—a change in the association of the corpuscles of the dissolved substance. Boyle tended to deemphasize chemical combination in favor of a mechanical alteration.[21]

Recipes for the Alkahest

The recipe for the alkahest was often kept secret. Some recipes, however, have been recovered—none of which lived up to the ideal of the theoretical alkahest. Paracelsus's recipe consisted of caustic lime (calcium hydroxide), alcohol, and carbonate of potash (potassium carbonate).[22]

Helmont claimed that knowledge of the recipe was granted by God, and therefore only revealed to those deemed worthy. He had dreams in which he believed the recipe had been revealed to him, only to find them inadequate. He then offered the use of other, inferior substances which he believed capable of similar tasks. Volatile salt of tartar (pyrotartaric acid or glutaric acid—$C_5H_8O_4$—produced by the body during the process of metabolism of certain amino acids) was considered both a substitute for the alkahest and a component of the alkahest. Helmont also referenced a 14th-century alchemical manuscript that discussed sal alkali—possibly a solution of caustic potash in alcohol—that was capable of dissolving many substances, and may have been an ingredient for his alkahest.[23]

Following a suggestion by van Helmont, George Starkey attempted to volatize salt of tartar, but without success. Starkey's term for the volatilized salt of tartar was *succedaneum*—not an equivalent but capable of performing some of the same feats.[24] Alkalies were known corrosives, so he reasoned that if salt of tartar could be volatized, its corpuscles would be reduced in size comparable to those of the alkahest (volatility and corpuscle size had been related in alchemy since the Middle Ages). The problem preparing a volatized salt of tartar (potassium carbonate) was that it was steadfastly non-volatile—characteristic of alkalies. Starkey pursued several processes, employing, separately, spirit of wine, spirit of vinegar, and sal ammoniac (ammonia chloride—ClH_4N), but without success. He did achieve, however, useful substances with these attempts. With the use of the spirit of wine to volatize salt of tartar, he produced the mendicant Balsamus Samech,[25] which was also known to Paracelsus and van Helmont. The trial using sal ammoniac resulted in a diuretic salt. Starkey reasoned that he could induce volatility by combining salt of tartar with volatile substances.[26] The volatilization of alkalies became a project unto itself for Starkey, who pursued it using plant oils such as olive oil.[27] In the end, he thought the answer was to be found in non-acidic and non-alkalic substances. Since acid saline liquors are destroyed by alkalies and urinous spirits, acidic substances could not be ingredients. Instead, he postulated that possible ingredients could include urinous spirits, spirit of alkalies, and sulphurous vegetable spirits. The secret ingredient of the alkahest, according to Starkey, was to be found within urine.[28]

Johann Rudolph Glauber (17th century alchemist, chymist, and apothecary who discovered sodium sulfate, a laxative, in 1625) believed that volatile niter (nitric acid) and fixed niter (potassium carbonate) constituted the alkahest, since they were capable of dissolving many substances. Glauber had discovered that saltpeter could be "fixed" by burning it with charcoal, which produced potassium carbonate (K_2CO_3), a caustic and cleansing salt. It was also common knowledge at that time that spirit of niter (nitric acid-HNO_3) could be produced by distilling saltpeter with fuller's earth (Fuller's earth consists primarily of hydrous aluminum silicates (clay minerals) of varying compositions). Thus

Glauber concluded that saltpeter was a two-fold substance—a "hermaphroditic salt," containing a volatile acid, which he referred to as volatile niter, and a solid caustic substance he called fixed niter (potassium carbonate). Since these two components of niter could dissolve various substances, Glauber believed he had discovered Helmont's alkahest.

While major figures such as van Helmont, Starkey, and Glauber pursued various organic and non-organic substances in their quests for the alkahest recipe, Frederick Clodius (German physician, 1625-after 1661, and an associate of Robert Boyle) advanced the Geberian school of alchemy in transmutational alchemy. Clodius maintained that quicksilver must be the starting point of the Philosopher's Stone and that the alkahest was to be made with the aid of common mercury, which acts upon salts and converts them into liquors.[29]

These are only a sampling of the recipes found in the alchemical texts, but represent the array of ideas and goals of the alchemical tradition: from transmutational alchemy—the "healing of the metals" to purify ores and ultimately produce gold and such arcanum as the Philosopher's Stone—to medical alchemy—the "healing of natural bodies" to cure incurable diseases and prolong life.

Conclusion

Though the alkahest, as envisioned by the alchemists, was never discovered, the pursuit of the "ignis gehenna" advanced chemical knowledge. New substances or improved understandings of the properties of known substances were discovered. Methods and techniques for carrying out chemical processes were improved as well as new ones created. Insights into chemical processes were gained. Pharmaceuticals, facilitating the treatment of diseases, were developed. The pursuit of the alkahest had vindicated the alchemists' and chymists' faith in the epistemology of separation.

CHAPTER 12

Notes

1. Quintessence is the Latinate name of the fifth element used by medieval alchemists for a medium similar or identical to that thought to make up the heavenly bodies.
2. Bruce T. Moran, *Distilling Knowledge: Alchemy, Chemistry, and the Scientific Revolution* (Cambridge, MA: Harvard University Press, 2005), 72.
3. Victor D. Boantza, "Reflections on Matter and Manner: Dulcos Reads Boyle, 1668-69," in *Chymists and Chymistry: Studies in the History of Alchemy and Early Modern Chemistry*, edited by Lawrence M. Principe (Sagamore Beach: Watson Publishing International, 2007), 183-184.
4. Guido Panzarasa, "Rediscovering pyrotartaric acid: a chemical interpretation of the volatile salt of tartar," *Bulletin for the History of Chemistry*. 40: 1–8.
5. Mark Haeffner, *Dictionary of Alchemy* (Aeon Books, 2004), 42-43.
6. Ana Maria Alfonso-Goldfarb; Marcia Helena Mendes Ferraz; and Piyo M. Rattansi, "Seventeenth-century 'treasure' found in royal society archives: The ludus helmontii and the stone disease," *Notes & Records: The Royal Society Journal of the History of Science*, 68 (3), 2014, 227–243.
7. The correct term is Philosopher's Stone, not the commonly used Philosopher's Stone. All original sources in various languages use the possessive plural, Philosopher's Stone: Stone of the Philosophers— see Lawrence M. Principe, *The Secrets of Alchemy*, (Chicago: The University of Chicago Press, 2013), 217n.
8. Principe, *The Secrets of Alchemy*, 134.
9. Ana Maria Alfonso-Goldfarb; Márcia Helena Mendes Ferraz; and Piyo M. Rattansi, "Lost Royal Society Documents on 'Alkahest' (Universal Solvent) Rediscovered". *Notes and Records of the Royal Society of London*, 64 (4), 2010.
10. Principe, *The Secrets of Alchemy*, 134; and Moran, 93.
11. William R. Newman and Lawrence M. Principe, *Alchemy Tried in the Fire: Starkey, Boyle, and the Fate of Helmontian Chymistry*, (Chicago: The University of Chicago Press, 2002), 138.
12. Helmont saw a resemblance between the "cold fire" of the alkahest and the "thick water" mentioned in the Book of Maccabees in the Old Testament.

13. Famously known for Boyle's Law, which expresses the inverse relationship that exists between pressure and volume of a gas.
14. Ana Maria Alfonso-Goldfarb, et al., "Lost Royal Society Documents on 'Alkahest' (Universal Solvent) Rediscovered."
15. Eirenaeus Philalethes, *The Secret of the Immortal Liquor Called Alkahest or Ignis-Aqua* (London, 1683), reprinted by The Alchemical Press, Edmonds, WA, 1984.
16. Principe, *The Secrets of Alchemy*, 134-135.
17. The "semina" was understood as an internal principle—a hidden and self-moved, internal efficient cause. The "semina" resided within physical bodies and were responsible for the specificity, uniqueness, transmutations, and development of those bodies. The "semina" worked radially, i.e., without physical contact. Thus it was the "semina," according to van Helmont, that transmuted primordial water into all other substances. See Newman and Principe, 62; and William Newman, *Gehennical Fire: The Lives of George Starkey, An American Alchemist in the Scientific Revolution*, (Chicago: University of Chicago Press, 1994), 143-146.
18. Paolo Porto, "Summus atque felicissimus salium": the medical relevance of the liquor Alkahest," *Bulletin of the History of Medicine*, Spring 2002, 76 (1):1–29.; and Newman and Principe, *Alchemy Tried in the Fire*, 138, 242–243, 249, 282, 286, and 292-296.
19. Porto, "Summus atque felicissimus salium": the medical relevance of the liquor Alkahest," 3-4; and Ana Maria Alfonso-Goldfarb, et al., "Lost Royal Society Documents on 'Alkahest' (Universal Solvent) Rediscovered."
20. Boyle learned of van Helmont's thought through George Starkey. See Newman and Principe, 295.
21. Newman and Principe, 292-296.
22. Porto.
23. John H. Leinhard, "No. 1569: Alkahest," Episode No. 1569: Alkahest (uh.edu).
24. Newman and Principe, 138.
25. A Paracelsian medicament prepared by digesting spirit of wine with salt of tartar; the salt (potassium carbonate) absorbs water from the spirit of wine (dilute ethanol) and dissolves itself into a liquid. Alchemical Glossary: The Chymistry of Isaac Newton Project, (indiana.edu).

26. This concept is further expounded upon in Robert Boyle's "Mechanical Origine of Volatility," in *Mechanical Origine of Qualities,* in *Works,* 8:432. See Newman and Principe, 293.
27. For a full treatment of Starkey's efforts, see Newman and Principe, *Alchemy Tried in the Fire,* 138-154.
28. Newman and Principe, *Alchemy Tried in the Fire,* 282-286.
29. Ibid, 242-249.

Part III

Concerning the Matter of Apparatus

Chapter 13

Alembics

Introduction

From archeological evidence, distillation devices may have been in use since 3500 BCE. Thus, by the last century BCE and first century CE, the construction and use of such vessels would have been well-known among alchemists. Only fragmented materials, however, have been found, and any conclusions must be cautionary. The evidence, dated to around 3500 BCE, is for four ceramic bowls each with a double rim, around which—it was postulated—lids would fit. A vapor could then condense on the inner surface of the lid and collect in the gutter. Other archeological evidence suggests that various types of apparatus were used for distillation, but, at best, all could only be described as proto-retorts or as simple stills since the evidence before the third century CE does not meet the definition of an alembic. For instance, remains dated between 90 BCE and 25 CE and Dioscorides's still (see the illustration below) could be conjectured to form some kind of a very early distillation device.[1] (Dioscorides (40-90 AD) also described a sort of still for the sublimation of cinnabar to make mercury. In this pre-alembic assembly, the cinnabar was heated on an iron saucer in a pan covered by a pot called an "ambix," on which the mercury vapor would condense.)[2]

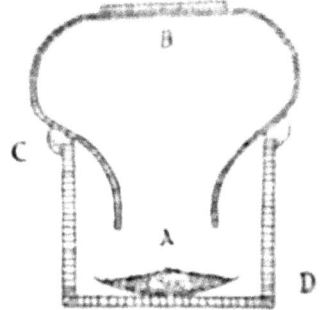

The Mercury-Still of Dioscorides from *Alchemists, Founders of Modern Chemistry*

Along with these early stills and distillation apparatus, the alchemists would have inherited a vast array of associated technological knowledge from the disciplines of metallurgy, ceramics, glass-making, brewing, dyeing and coloring, weaving, and the preparation of drugs, poisons, and cosmetics.[3] which would have led to numerous innovations in their distillation devices.

An example of an alembic made of devitrified glass from the Near East[4]

The etymology of the word alembic traces back to the Arabic, *al-inbiq* (romanized), which originated from the Greek ἄμβιξ (romanized: *ambix*),[5] meaning cup or beaker. The alembic was one part of a distilling apparatus or still, which consisted of three parts—the cucurbit, the head or cap (alembic), and the receiver or container. The cucurbit (Arabic: *kar'a*; Greek: βῖκος, *bîkos*) was the still pot containing the liquid to be distilled, which was heated by a flame. The alembic was placed on top of and fitted over the mouth of the cucurbit, which allowed it to receive the vapors coming off the heated substance. A downward-sloping "tube" or solen (Gk: σωλήν) was attached to the alembic, allowing the condensed vapor to flow into a receiver or container (known in Arabic as *qābila*; or in Greek as ἄγγος, *angos* or φιάλη, *phialē*). In other words, the liquid in the cucurbit was heated or boiled to produce a vapor that rose into the alembic, where it cooled by contact with the walls, condensed, and then ran down the tube into a receiver. Typically, the principal fuel used to heat the cucurbit and its contents was charcoal, but other materials–wood, peat rushes, oil, wax, pitch, coal, and dried dung of horses, cattle, or camels–could be used as well. The alembic, then, had a specific and significant role in the distillation process. In later alchemical literature, however, the term alembic took on a broader meaning; it would often be applied to the cucurbit and head or cap as a whole.[6]

Alembic Symbols from *Alchemical Symbols*[7]

CHAPTER 13

The alchemists were innovative when it came to their instruments, adapting their apparatus and equipment to fit specific operations such as calcination, sublimation, fusion, crystallization, and distillation.[8] They were especially innovative in their methods of distillation. This creativity can be seen, especially, in the variety of descriptions they used to explain that process, as can be seen in what follows.

Cucurbit symbols from *Alchemical Symbols*[9]

In the early literature, two women are credited with being the first among the Greek alchemists to describe their distillation apparatus: Maria the Prophetess (also known as Maria the Jewess, who, according to Zosimos, lived between the 1st and 3rd centuries CE in Alexandria) and Cleopatra, the Alchemist (to be distinguished from the more famous Queen Cleopatra VII—the supposed lover of Mark Antony—and Cleopatra the Physician, who supposedly was a contemporary of Cleopatra the Alchemist). They are considered to be the first alchemists to describe their alembics.

Maria the Prophetess (c. 1st to 3rd century CE)

A depiction of Maria the Jewess from Michael Maier's book *Symbola Aurea Mensae Duodecim Nationum* (1617)

Some of their alembics—or still-heads—were provided with two or three spouts, and were then known as dibikos or tribikos, respectively. Zosimos attributed the innovation of the tribikos to Maria the Prophetess, who, he believed, was a practical alchemist who knew her way around her kitchen. He quotes her at length when he describes her instructions for making a tribikos:

> 'Make,' she says, 'three tubes from ductile bronze, thin metal in thickness little more than that of a frying pan for cakes, in length a cubit and a half. Make three such tubes and make one of the diameter of about 3 inches adjusted to the opening of the copper still-head. Let the three tubes have openings adapted to their little receivers; let there be a little nail for the thumb tube so that the two finger tubes may be adapted to the two hands from the sides. Near the edge of the copper still-head are three holes adjusted and let it be soldered to fit closely to that part which carries the vapor upwards in the contrary direction. And placing the still-head on the earthenware flask containing the sulphur and having luted the joints with flour paste, put large glass flasks on the ends of the tubes, so thick that they will not break with the heat of the water carried up the middle.[10]

A particularly innovative device used for the gentle heating of a still was also accredited to Maria the Prophetess. This "water-bath" still bears her name: the bain-marie (Mary's bath).[11] The bain-marie limited the maximum temperature of the cucurbit and its contents to the boiling point of the liquid of the bath. The device was essentially a double boiler, used for distillations in which gentle and gradual heating was necessary.

Maria's bath, from
Coelum philosophorum,
Philipo Ulstadio Patricio, 1528[12]

CHAPTER 13

Another distillation device attributed to Maria was the *kerotakis*. In antiquity, painters often used an encaustic process in which the pigments were mixed in melted wax and then applied with a brush. A metallic plate or palette (triangular in shape, perhaps similar to a bricklayer's trowel), known as the *kerotakis*, was used to hold the paints which were then kept fluid by resting the palette over a charcoalburner.

In alchemical operations, the *kerotakis* was placed inside a cylindrical or spherical container–closed at the lower end–which would contain mercury, sulfur, arsenic, or some other substance that would partly or entirely vaporize on heating. The metal to be exposed to the vapors, in the form of a foil or powder, was placed on the *kerotakis* (later, the name would come to denote the entire apparatus). The opening at the top of the container was closed with a hemispherical lid. When the volatile substance was heated, the vapors would, at least in part, react with the metal, and that which did not react would condense on the lid with the condensate flowing back into the container—in effect providing a continuous reflux action.[13] Maria may have used this apparatus in the treatment of copper and lead with arsenical or sulfurous vapors.

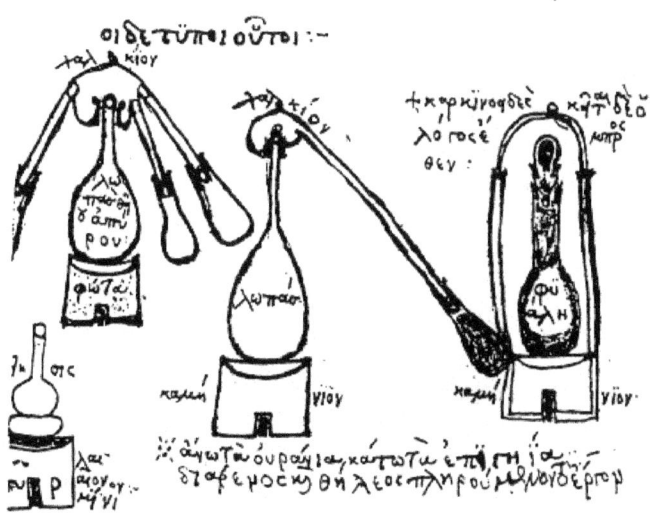

Chemical apparatus figured in MS. Parisinus 2327. From F. Sherwood Taylor, "The Origins of Greek Alchemy." (These drawings, the earliest illustrations available, were copied in the 10th or 15th century from the works of Greek alchemists, notably from the writings of Zosimos. The illustrations show a flask, an alembic, a condenser, a receiver, and a form of a reflux apparatus. The alembic in the upper left is a tribikos.)[14]

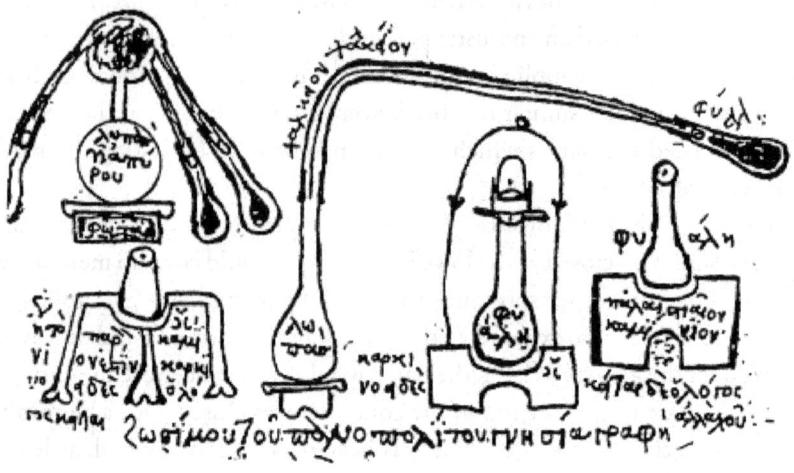

Chemical apparatus also figured in MS. Parisinous 2327. From F. Sherwood Taylor, "The Origins of Greek Alchemy." (This illustration shows a distillation apparatus, a tripod, a reflux apparatus, a hands-breath furnace, and a water-bath or ash-bath.)

Though Zosimos made use of the dibikos, the tribikos, the *kerotakis*, and other such apparatus, he did not devise them himself. He and later commentators[15] regarded the works of the Jewish alchemists, especially those of Maria the Prophetess, as the source of this information. Maria's works have all been lost, but Zosimos gave her credit for devising not only the instruments described above, but several kinds of furnaces as well.

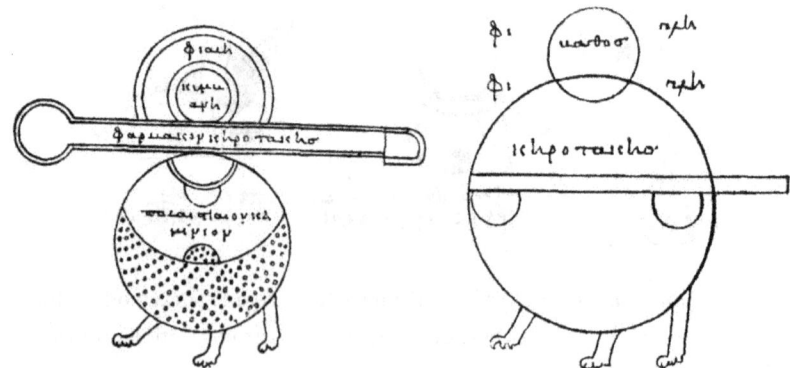

Drawing of the Bain Marie from Marcellin P.E. Bertholot and Charles E. Ruelle, *Collection des anciens alchemistes grecs* (Paris, 1888)[16]

CHAPTER 13

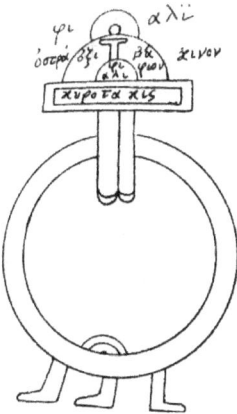

Drawing of a *kerotakis* from Marcellin P.E. Bertholot and Charles E. Ruelle, *Collection des anciens alchemistes grecs* (Paris, 1888)[17]

Cleopatra the Alchemist (c. 3rd century CE)

A Greek alchemist, author, and philosopher, Cleopatra the Alchemist[18] is believed to have been active in Alexandria. She was associated with the school of alchemy typified by Maria the Prophetess, which focused on the processes of distillation and sublimation and their associated apparatus. Along with three other women—Maria the Jewess, Medera, and Paphnutia[19]—she has been credited with devising a formula for making the Philosopher's Stone. Her most noted work was the *Chrysopoeia* (Gk: gold-making), in which appeared a drawing of a dibikos (Her other work was "On Weights and Measures").

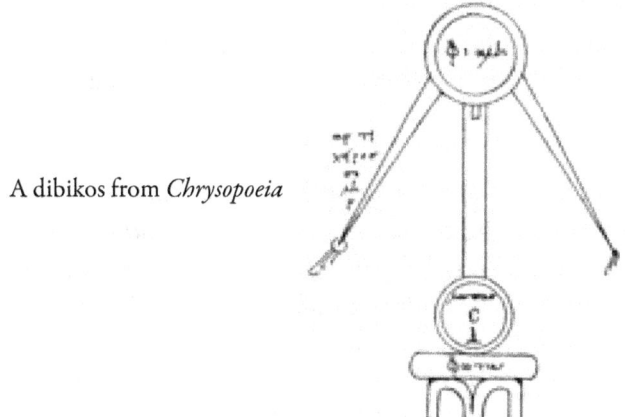

A dibikos from *Chrysopoeia*

Some writers considered her to have invented the alembic.[20] But whether she—or Maria the Prophetess— holds that distinction remains one of the unanswerable questions in alchemical history.

Zosimos (c. 290-330 CE)

Born in Panopolis (present-day Akhmin in the south of Roman Egypt) and known by the Latin name *Zosimus Alchemista*, (Zosimus the Alchemist), Zosimos was a Greco-Egyptian alchemist and Gnostic mystic. Along with his sister Theosebia, he wrote one of the oldest known books on alchemy, entitled *Cheirokmeta*, (Gk. "things made by hand" or "manipulations"), a chemical encyclopedia of 28 books.[21] As mentioned above, Zosimos did not devise alembics himself; instead, he utilized and properly credited the work of Maria the Prophetess. In particular, he made use of the *kerotakis* in his experiments on the actions of vapors on solids. The craftsmen of antiquity knew that the vapors released by cadmia (a zinc-containing material) could color copper golden (a result of making brass, an alloy of zinc and copper) and that the vapors of mercury and arsenic would whiten copper to a silvery color. It may have been this knowledge that led Zosimos to investigate processes that could result in transmutations of base metals. The *kerotakis* would have been ideal for these pursuits, since it was designed to enable condensates of various materials to react with metals or other substances.[22]

Al-Razi ((Abu Bakr Muhammad ibn Zakariya al-Razi, c. 865-923 CE)

In his *Book of Secrets*, al-Razi included detailed instructions regarding the equipment he used in his distillation processes. The equipment he used included the cucurbit, the alembic with beak and receiver, the vessels and blind alembic, the stove, the beakers, the flasks, the vials, the grinding plates and graters, the oven, the braziers, the self-ventilating oven, the containers, and round molds. According to al-Razi:

> . . . *the cucurbit, the alembic with the beak and the receivers are suitable for distilling liquids. The secret of this is that the vessel must be large and thick-walled, without fissures in the bottom, and that there must be no blisters in the walls, and the alembic must fit on it tightly. The kettle, in which the alembic is placed, should be shaped like a cooking pot, and the vessel must be immersed in the water (of the kettle) up to the highest level of the material that it contains.*

> *Furthermore, a large cauldron of boiling water must be ready on the stove to fill the kettle (or water bath) when (its water level) is reduced. And watch out lest the vessel touch cold water and secure the vessel so that it cannot move, and that its bottom does not touch the bottom of the kettle, or it will break.*

He also provided several options for heating the distillation apparatus. He cautioned, however, that it was necessary to tightly secure the end of the tube and the mouth of the receiver so that the smoke from the fire would not enter and contaminate the apparatus and its contents:

> *The vessels can also be placed in a kettle with (sifted) ashes and heated from beneath. That is the best method for beginners.*
>
> *One can also place a large brick on the bottom of the stove where the bottom of the kettle stands, and put ashes on it and stand the kettle on it and pack sifted ashes around the sides of the kettle. Then heat it; however, it is necessary for you to secure the end of the tube and the mouth of the receiver (make it tight) so that the smoke of the fire does not enter it and the air pollutes the contents.*[23]

In addition, al-Razi distinguished four types of alembics. One alembic had a wide spout for distilling "... the blackness from calcined substances ..." and was suited for the sublimation of sal ammoniac. Another alembic, which did not have an especially wide spout, was meant for the "... distillation of essences ..."—possibly referring to impurities and colors—or for sublimation. The third type was an alembic with a spout that was comparatively narrower than the first two and was used for "... distilling stones at the beginning of the work. ..." Finally, the fourth type included an alembic with a very narrow spout and was suited for the evaporation and purification of liquids.

The cucurbit with the "blind alembic" was used for dissolving spirits (strong solvents) and softened metals. He described this apparatus as an alembic with a gutter without a spout, into which the substance to be dissolved was placed. The "sharp water" (a strong solvent) was then placed in the cucurbit. The alembic was then set on top of the gutter with the connection sealed, and the whole assembly positioned in a kettle of water (water bath). This apparatus, according to al-Razi, was only suitable for dissolving substances. (From al-Razi's description, the water-bath must have been widely known and used.)

The "blind" mentioned by al-Razi was a fitted beaker set on top of a cucurbit, in which substances to be dissolved were placed. The assembly, then, was hung in a fitted oven, under which was set a burning lamp, coal, or hot ashes. In his description, he further admonished that one must ensure that the fire did not go out or, if ashes were used, that the ashes did not get cold before the substance in question had dissolved and solidified.

After he described the construction of various distillation apparatus, al-Razi pointed out that the equipment was to be made out of glass, potter's clay, stone, iron, or crucible clay. All of these were to be coated with "artist's clay," which was made from "...pure red or white viscous clay, free from stones...." This mixture was to be spread out in a clean place, and water was to be sprinkled on it several times until it was saturated and dissolved and "...its grains so fluid that the hand no longer feels them." The remainder of the recipe involved a number of specific steps and detailed instructions, which indicated that al-Razi considered this material as a lining critical to the success of the distillation operation:

> *Let it [the mixture referenced above] stand there until it is dry (again), then pulverize it as the jar makers do, and sift it through a flour sieve and grind it with a mortar and then saturate it with water, in which rice bran, from which flour is made, was softened. Then knead (the clay) thoroughly and let the dough soften a day and a night. You then take clean rice and sift it through a large hair sieve, so that all the dust falls out of it (again?), mix it with an equal amount of the softened potter's clay, adding to each pound (ratl) the weight of ten dirhams of table salt, and the same amount of rice, and a third pound of ground and silk-sifted pot shards, and a handful of animal air, cut up as fine as possible. Let it stand three days and use it, for it is the best artist's clay that there is. And success comes from God.*[24]

Since al-Razi included aludels in his list of equipment, that particular type of apparatus must also be considered. In the *Book of Secrets*, he treated the equipment used for distillation and sublimation processes within the same section, "The Equipment for Handling Nonmetals." According to al-Razi, the aludel was for the sublimation of dry matter, while the alembic was for the distillation of "waters."[25] Sublimation occurred in clay vessels which were placed in a saucer of potter's clay on a stove and then heated with a gentle fire. The aludel was a series of pear-shaped earthen tubes (vials) or pots, which were open at both ends so they could fit one over the other, diminishing toward the top. The lowest tube or vial was adapted to a pot, which was placed in a furnace. The tubes or vials, then, forced the substance upwards into the upper section where it was volatilized and confined.[26]

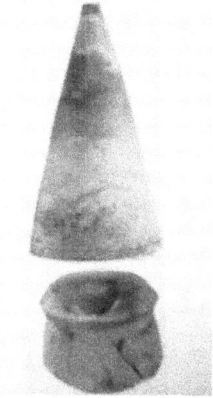

An earthenware aludel and subliming pot[27]

Al-Razi described the construction of his aludel in the following manner:

> ... take a vessel in the shape of a large pot one ell long and two fists wide. Then lay it on a flat surface and scatter (one and a half fists) sifted ashes all around it. Then remove it again and place a cover made of artist's clay on the sifted ashes around the kettle, let it dry, and lift it (out of the mold). Then ... coat the outer surface and smooth it over with (a mixture of) white lead and egg white, and coat it a second time and make a gutter around its edge and leave a place in it open, so that the sublimated substances can be collected ... now let it dry, then ... turn the upper part of the aludel with the opening beneath and coat evenly it with clay from medium grain, not too big and not too fine. Then set the cover on the kettle and seal the joint with clay on all sides. Also make a continuous wing under the cover, so that the fire does not come in contact with what is on the cover....[28]

Later, in the "Second Chapter on the Sublimation of Mercury," al-Razi–according to the translation of the *Book of Secrets* found in Taylor's *The Alchemy of Al-Razi*–identified and described the entire aludel as an alembic:

ALCHEMICAL MATTERS

And the aludel is an alembic of clay or glass, with a wide spout. You use it to distill everything that has moisture in it. You set a bowl on top of it, or set a well-fitting lid on top of the kettle. There is a hole in the lid the size that the head of a strong needle can pass through....[29]

(H. E. Stapleton, et. al., however, understood al-Razi's description as identifying the alembic as the head of the aludel. This would indicate the possibility that the terms aludel and alembic were either confused or conflated.)[30]

Symbol for an alembic from *Alchemical Symbols* by Philip Wheeler[31]

Avicenna (Ibn Sina, 980-1037 CE)

Avicenna, a philosopher and physician, described the uses of alembics in his *The Treatise of the Most Excellent of the Moderns* in chapter VII entitled "Extraction of the Tincture of Sulphur":

Another method is to place under it a lamp with a small flame so that the liquid may not boil and the sulphur burn. With either method, the red color passes out into the water; but it (the mixture) should be shaken several times each day. After the red water is drained off, fresh water is poured onto the sulphur (and the process repeated), until it no longer turns red. Then all these waters are mixed together and distilled in a narrow alembic, when the tincture will (finally) remain in the cucurbit (qar'a) close to the anbiq. If any trace of redness remains in the water (that passes over), the distillation is repeated until the tincture has been completely separated.

A comparison to the *Kitab al-Asrar* (Book of Secrets), which described the construction of an alembic, provides an understanding of the workings of the alembic mentioned here. The circular opening in the base of the alembic had the same diameter as the outside top of the cucurbit, and the joint was closed with clay. The liquid portion of the mixture was then passed down the spout that emerged from the top of the alembic and into a glass receiver. The sublimed tincture, in the above description, was then collected as a solid in a circular channel made of clay inside the cucurbit near its top.[32]

Pseudo-Geber (Paul of Taranto, 1290-1330)

From the chapter "Of Distillation, and its Causes, and of Three kinds of the same, viz., by Alembeck, by a Descensory (a vessel used to extract oils), and by Filter" in his work, *Summa Perfectionis*, Geber described distillation as:

> ... an Elevation of aqueous Vapors in their Vessel. And Distillation is diversified. For Distillations are by Fire, and some without Fire. Those made by Fire are of two kinds; one, which is by Elevation into the Alembeck; and the other by Chymical Descensory, by mediation of which the Oyl of Vegetables is extracted. The Cause why Distillation was invented, and the general Cause of the Invention of every Distillation, is the Purification of Liquid Matter from its turbulent Feces (impurities)....[33]

This was followed by a description of distillation methods employing an alembic:

> Now we will shew you the Methods of Distillations, with their Causes. Therefore, of that which is made by Ascent, there is a twofold Way or Method. For one is performed in an Earthen Pan full of Ashes; but the other with Water in its Vessel, with Hay or Wool, orderly so disposed, that the Cucurbit, or Distillatory Alembeck, may not be broken before the Work is brought to Perfection. That which is made by Ashes is performed with a greater, stronger, and more acute Fire. For Water admits not the Acuity of Ignition, as Ashes doth...Therefore, more subtle Separation is made by Distillation in Water than by Distilling in Ashes....[34]

After delineating the causes for the two methods, Pseudo-Geber described the equipment used in the methods of "Ascent." To distill in ashes, a strong earthenware pan, layered with ashes and attached to a vented furnace, was employed. The distillery or cucurbit was placed on the bed of ashes with a thickness of one finger and covered with the same ashes almost to the neck of the alembic. Next, the matter to be treated was placed into the cucurbit, which would then be covered by the alembic. The neck of the alembic enclosed the neck of the cucurbit up to the curved channel of the alembic, and the joint of two necks luted (This was to prevent the distilled vapor from escaping). Both the cucurbit and the alembic were made of glass.

The second method of "Ascent" involved an iron or brass vessel, instead of a pan, attached to a vented furnace. At the bottom of the vessel, a bed of hay or wool was laid on which the cucurbit with its alembic sat. The hay or wool was a protective layer to prevent direct exposure to concentrated heat, which would cause the cucurbit to break. The hay or wool, then, was placed all around the cucurbit up to the neck of the alembic, as was done for the distillation in ashes. Next, a grid of sticks was laid across the cucurbit and alembic, on which stones were placed. This kept the cucurbit and alembic firmly in place, which prevented their movement when water was added to the pot.[35] The description of Pseudo-Geber's set-up was similar to the descriptions of a water-bath or Bain-marie offered by other alchemists.

Distillation apparatus and furnaces from the *Book of Furnaces*, attributed to Geber[36]

CHAPTER 13

Vannoccio Biringuccio (c. 1480 – c. 1539)

Vannoccio Biringuccio was born in Siena in 1480. He was the first metallurgist to attempt to incorporate the entire field of metallurgy in his work *Pirotechnia*, consisting of ten books, first published in 1540. In that work, he covered the process of distillation and the application of alembics in detail, devoting chapter two of the ninth book to distillation and sublimation. According to Biringuccio, the methods and equipment employed in distillation processes were common.

Nevertheless, Biringuccio introduced new terms in describing alembics and associated equipment. He referred to one of the apparatus as a bell, since, according to him, it resembled the shape of a bell. The bell consisted of a shell which contained the material and a cover on top—both made of lead, glazed earthenware, or tinned copper. From this description and the figure below, taken from *Pirotechnia,* Biringuccio's bell was the same as a variant of an alembic (later, in his description, he referred to the entire assembly of the bell as an alembic. See the discussion of John French below), which was also commonly used by apothecaries and others who had materials of "... great moistness ... [or] ... a quantity of material to be distilled." Biringuccio continued his description by adding that the cover had a wide bottom and a circular hollow retainer similar to a channel attached all around the inside of the cover (note the similarity of this description to the archeological evidence of lids with gutters mentioned earlier). This channel, then, collected the condensed vapor that "... the smoke breathes up into the air of the bell because of the heat of the fire...." The condensate would run down along the inside surface of the dome of the cover, which was then carried away by a retainer (a long hollow tube or solen) to a receiver. To prevent any loss of the vapor, the components were to be fitted as exactly as possible, with the joints, parts, and rims conformed to the mouth of the shell that contained the material.

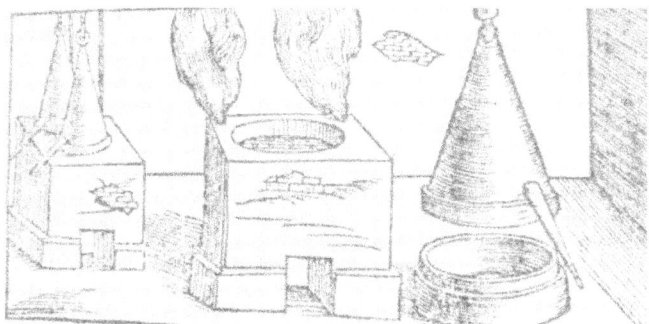

Metal distilling bells and their furnaces from *Pirotechnia*

283

Later, in describing the distillation process used for "...extracting waters from drier and more resistant things...," his list of required apparatus included cucurbits of glass, clay, or tinned copper. For making acids, he recommended making the cucurbits of glass, which was to be as clear and uniform as possible with no bubbles or indentations. Some of the cucurbits were to have wide mouths shaped like "urinals." On top of the cucurbits, glass alembics were to be placed.[37] In regard to the luting, Biringuccio advised using *lutum sapientiae*, which was to be spread:

> ... *up to three dita or less from where the neck narrows. This lute must be spread out well all over to the thickness of two threads or a little more. Then cover and make them strong with this well-made plaster, and finally dry it, taking care that there are no clefts or cracks such as clays are often prone to make, though they be well prepared. For greater security, choose among these a clay that is lean and then mix it with at least a fourth part or more of wool-cloth clippings and about an eighth part of wash ashes, a quarter of mule or horse dung, or that of some other animal whose dung is dry. These things are all mixed together and beaten with an iron rod. This is the composition that the alchemists call lutum sapientiae, with which the bottoms of the cucurbits that you wish to use are plastered and strengthened.*[38]

Cucurbits, alembics, and receivers arranged in the furnace for distillation, from *Pirotechnia*

Biringuccio classified distillation processes into two methods: one with heat and "dryness" and the other with heat and "moisture." The first method entailed placing earthenware or ashes between the cucurbit and the source of the heat. The second method involved placing the cucurbit with its alembic in a kettle of boiling water (see the figure below) or some other suitable vessel.[39]

Distillation with a cucurbit and alembic in a water bath, from *Pirotechnia*

If it were desired to make a more "penetrating or subtle water" (e.g., a stronger acid), then Biringuccio suggested using a Pelican (see below). The more subtle distillate was achieved by placing the condensate from a previous operation into other vessels such as the Pelican for circulation and distillation again. The re-distillation was performed many times such that the substance of interest was reduced "... almost to the subtlety of smoke, so that when the vessel containing it is opened it goes away in the air. ..." This process of re-distillation was also known as cohobation, a technique in which a liquid was distilled off of a substance, and then poured back over the residue and distilled off again. This procedure would often be repeated many times (A good example in the alchemical literature of this process was that presented in the Third Key of Basil Valentine's work *The Twelve Keys*).[40]

Pelicans and other vessels for reflux distillation, from *Pirotechnia*[41]

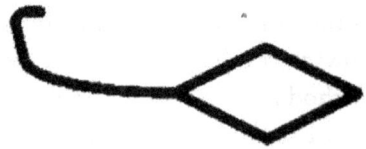

Cohobation symbol from *Alchemical Symbols*[42]

Other interesting variations on distillation apparatus included the worm condenser and the rectifying column (both are shown in the figure below). The worm condenser was a tube made of tinned copper that extended from the alembic in which the wine was put. At the end of this long tube, at four or six "braccia,"[43] was a vat made of either copper or wood, in which the tube continued with several snakelike bends.

The rectifying column extended from the top of the vessel, entered a glass hood from which the *aqua vitae* came out, and passed into a receiver. The vessel was placed in a small furnace and filled with wine through a tube attached to the side opposite from that to which a spigot was attached. A small vat was present at the bottom end of the worm-shaped section of the tube and contained cold water for cooling. Finally, a receiver was attached to the alembic via the spout at the top.[44]

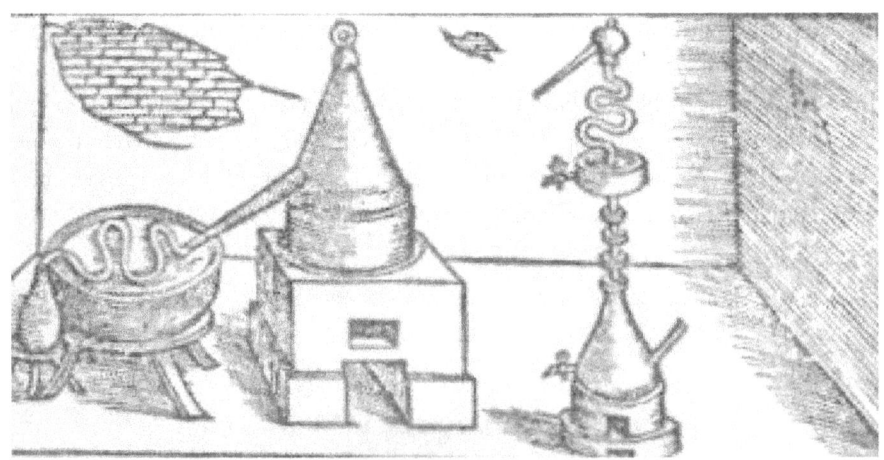

Distillation of alcohol in a bell with a worm condenser and with a rectifying column, from *Pirotechnia*

Another variation of the distillation apparatus involved the distillation of "oils." According to Biringuccio, the most common means was exhalation,[45] not only with the force of fire, but with the appropriate apparatus. His recommendation was to use cucurbits with "...bent necks and the rest shaped like the sack of a bagpipe, with the neck turned downwards rather than level...." These were luted all over their bodies and up the bend of the neck.

Retort[46] [cucurbit w/bent neck] and furnace for the distillation of oils[47]

Besides the distillation applications and apparatus discussed above, Biringuccio covered several other applications and variations of distillation apparatus, detailing their construction, their luting, and their operation.

John French (1616-1657)

Similar to Biringuccio, John French, an English physician, also wrote detailed descriptions of distillation processes and their associated equipment in *The Art of Distillation,* which provided three principal definitions of distillation:

1. *Distillation is a certain art of extracting the liquor, or the humid part of things by virtue of heat (as the matter shall require) being first resolved into a vapor, and then condensed again by cold.*
2. *Distillation is the art of extracting the spiritual, and essential humidity from the phlegmatic, or of the phlegmatic, from the spiritual.*
3. *Distillation is the changing of gross thick bodies into a thinner, and liquid substance, or separation of the pure liquor from the impure feces.* [48]

To carry out the distillation operations, French offered various materials and forms that could be employed. He asserted that the vessels could be made of copper, iron, or tin, which were better than lead since lead would turn the "...Liquors into a white and milky substance, besides the malignity they give to them...." The best materials, however, were Jug-metal, Potters-metal glazed, or glass. These he affirmed could be used without fear of breaking or melting. He also indicated that some vessels could be made of silver, but "...they are very chargeable."[49]

For the luting material used to coat glasses, sealing closures, and stopping glasses, he recommended the following:

> *Take of loam and sand tempered with salt water (which keeps it from cleaving), to these add the Caput Mortuum of vitriol, or Aqua Fortis, and scaling of iron, and temper them well together, and this serves to coat retorts or any glass vessels that must endure a most strong fire, and will never fail if well made. Some add flax, beaten glass, and pots, and flints...*[50]

French also provide detailed drawings illustrating the construction of the distillation apparatus.

A cold still

This illustration clearly shows the position of the groove and the attached spout or solenoid. The alembic, then, sat over the cucurbit which formed the base of the still and contained the material in question. French referred to this apparatus as a cold still since it was used to capture a condensate with the odor of herbs and flowers with or without the presence of heat. The use of the still was described as a simple operation:

> *... note that this kind of water is but the flegm of the vegetable or odor in it; only roses and mints and two or three more have an odor, but all besides as little virtue as common distilled water.*
>
> *To make waters in a cold still that shall have the full smell and virtue of the vegetable. Take what herbs, flowers, or roots you please (so that they be green), bruise them and mix them with some leaven, and let them stand close covered for four or five days: then distill them after the manner aforesaid.*[51]

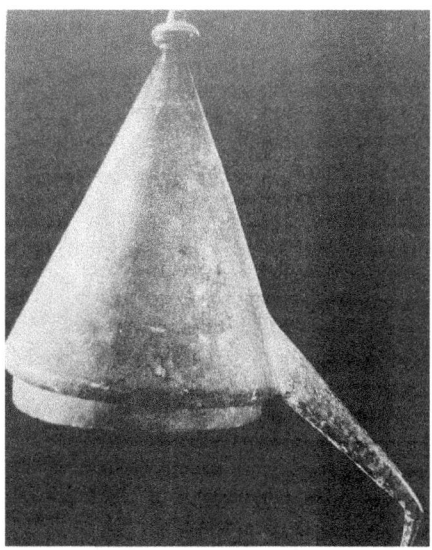

"Rosenhut" alembic made of pewter of unknown origin. (Compare the drawing above provided by the French.)[52]

If a stronger smell was desired, the French recommended the application of a gentle heat to the cucurbit containing the herb or flower.

ALCHEMICAL MATTERS

The form of an alembic[53]

In the above design "A" signified the cucurbit or vessel which, according to French, must be made of copper and would contain the matter to be distilled. It had to be set over a "naked fire." "B" was the belly that was attached to the neck so that the neck would properly fit the mouth of the vessel. (The mouths of the upper and lower vessels could be so fitted that this belly was not necessary.) The neck, "C," of the upper vessel allowed the "spirit or waters" to cool and then condense in the head, "D." "E" encompassed the head, into which cold water was continually poured. The long receiver, "F," would capture the condensate, and "G" was the top or cock which would let out the water as it became hot.[54]

The form of a Pelican[55]

CHAPTER 13

Pelicans were used for a process called circulation, a long-term reflux distillation process. French recommended the pelican for making "the Magistry of Wine:"

> *To make the Magistry of Wine which will be one of the greatest Cordials and most odoriferous Liquor in the World. Take good old rich Canary Wine, put it into a glass vessel that it may fill the third part thereof, and nip it up and set it in a continual heat of horse dung for the space of four months. Then in frosty weather set it forth into the coldest place of air you can for the space of a month that it may be congealed. And so the cold will drive in the true spirit of the wine into the center thereof and separate it perfectly from its phlegm. That which is congealed cast away. But that which is not congealed esteem as the true spirit of the wine. Circulate this in a Pelican with a moderate heat for the space of a month, and you will have the true Magistry or spirit of wine, which, as it is most cordial, so also most balsamical, exceeding all balsams for the cure of wounds.*[56]

The Pelican had two side tubes that would recycle the condensed vapor which collected on the surface of the head and ran down inside the tubes to the base for re-distillation. The problem with the pelican was that it would have been difficult to fill and seal. Also, very effective luting would have been necessary to avoid the loss of any vapor.[57]

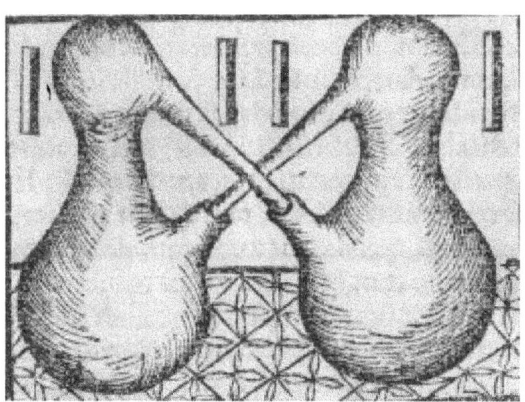

Another version of the Pelican.
Drawing from *Coelum philosophophorum* by Philipo Ulstadio Patricio.

In addition to illustrating the construction of alembics and stills, French also depicted how his vessels were connected to furnaces. The set-up below was used to distill liquors with the steam from boiling water.

A Furnace with his Vessels from
The Art of Distillation by John French[58]

The alembic, "A," and the body of the alembic, "B," were placed in a brass vessel, "C," which was made to contain the alembic body. Sawdust was placed in "C," surrounding the alembic body so the heat of the vapor would be retained more effectively and for a longer period of time. The sawdust also served as a barrier between "B" and "C" in order to prevent cracking from the direct contact of the two components. "D" indicated the brass vessel, containing the water that was placed in the furnace. To the furnace, "E," was attached a funnel, "F," by which water was added, as needed, to replace the water that had been evaporated. The condensed vapor was then captured by the receiver, "G."[59]

French also sketched several of his versions of the bain-marie, which are shown below.

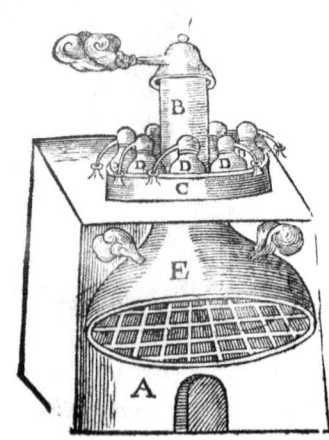

The delineation of a Balneum Marie from
The Art of Distillation by John French[60]

CHAPTER 13

This version illustrated the use of two furnaces, "A" and "B." In the figure, "A" had a hole from which the ashes were taken. "B," made of brass, was set inside the other furnace so the contained water and ashes would be heated more effectively. The alembics, "D," were set in the kettle, "C," which contained water, ashes, or sand. The spouts of the alembics transferred the condensed vapor into receivers. "E" represented the bottom of the second furnace, "B," which contained the fire.[61]

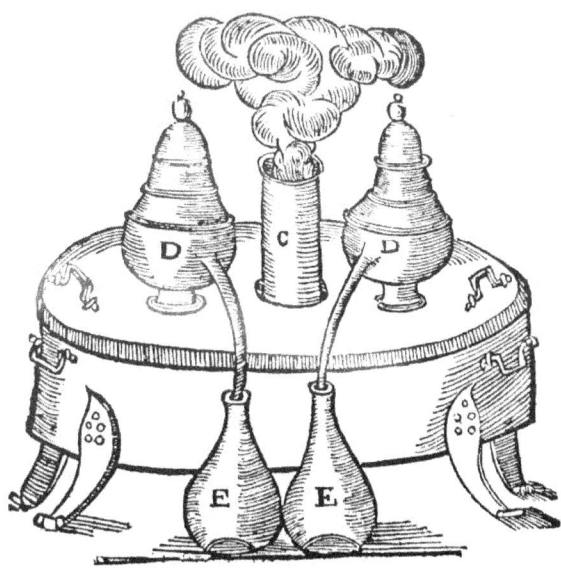

Another version of the Balneum Marie from *The Art of Distillation* by John French[62]

The kettle, made of brass, was filled with water while the lid was perforated in two places to allow the alembics, "D," to be affixed and to allow the passage of the heat of the water vapor. In this case, a chimney, "C," was utilized to relieve the resulting pressure. The distillate or distillates (in this option and the one above, more than one substance could be distilled.) was then transferred to the receivers, "E."

For his time, French's *The Art of Distillation* was considered the definitive work on distillation. In addition to being a handbook on the construction and proper use of alembics, cucurbits, furnaces, and other related equipment, his book was also an extensive source of recipes for the distillation of various materials and substances.

Conclusion

From ancient times, people sought to understand and utilize the substances of the natural world. Their endeavors revealed that materials could be broken down or changed into other materials by the application of heat through various means. One of the methods discovered was distillation—the employment of fire to drive off vapors from various substances and then condense the vapors into liquids, which could be used in the creation of various products—useful substances such as medicines, utensils, weapons, tools, cosmetics, wine, etc. From the ingenuity of the alchemists, various distillation apparatus, associated equipment, furnaces, and methods were developed—each designed to accommodate the unique conditions of the substances and the consequent requirements for their vaporization—and the alembic was a crucial element in the implementation of distillatory operations. In addition to facilitating the utilization of materials from the natural world, distillation enabled the growth of technological and chemical knowledge, becoming one of the most significant means that accelerated the evolution of alchemy into modern chemistry.

CHAPTER 13

Notes

1. R. G. W. Anderson, "The Archeology of Chemistry," in *Instruments and Experimentation in the History of Chemistry*, Frederic L. Holmes and Trevor H. Levere, Editors, (Cambridge, Massachusetts: The MIT Press, 2000), 5-22.
2. F. Sherwood Taylor, *Alchemists, Founders of Modern Chemistry, Kessinger's Legacy Reprints,* (Whitefish, MT: Kessinger Publishing, 2010), 37-38.
3. E.J. Holmyard, *Alchemy*, (Garden City, NY: Dover, 1990), 43.
4. Anderson, 11.
5. Merriam-Webster.com, "Alembic," https://www.merriam-webster.com/dictionary/alembic. Retrieved 30 July 2023.
6. Holmyard, 46-48.
7. Philip Wheeler, *Alchemical Symbols, Fourth Edition, R.A.M.S. Library of Alchemy, Vo. 21*, (Kansas City, MO: R.A.M.S. Publishing Co., 2018), 16.
8. E.J. Holmyard, 45. Also, there were many other operations listed by various names throughout the alchemical literature.
9. Wheeler, 34.
10. F. Sherwood Taylor, "The Origins of Greek Alchemy," *Ambix*, 1:1, 30-47. See also E.J. Holmyard, Alchemy (Garden City, NY: Dover, 1990), 47-49.
11. Holmyard, 48. The term was actually introduced by Arnold of Villanova in the 14th century. The bain-marie was also used in cooking food.
12. A nobleman from Nuremberg, who taught medicine at the academy in Fribourg at the beginning of the sixteenth century. He is known for his work *Coelum philosophorum seu de secretis naturae liber* (Fribourg, 1525), a major work of early modern distillation technology. *The Coelum philosophorum* contains extracts from Arnold of Villanova, Ramon Llull, Albertus Magnus, and John of Rupescissa. Atkinson, Edward R., and Arthur H. Hughes. "The Coelum Philosophorum of Philipp Ulstad." *Journal of Chemical Education*, 16 (1939): 103–107
13. Holmyard, 49. See also F. Sherwood Taylor, "The Origins of Greek Alchemy."
14. One of several manuscripts containing the works of the early Greek alchemists (before 1000AD). See F. Sherwood Taylor, "The Origins of Greek Alchemy."

15. F. Sherwood Taylor, "The Origins of Greek Alchemy."
16. Raphael Patai, *The Jewish Alchemists: A History and Source Book* (Princeton, NJ: Princeton University Press, 1994), 60-63.
17. Ibid.
18. Her name may be a pseudonym for an unknown author or group of authors. She was first mentioned in the "Dialogue of Cleopatra" from the *Book of Komarios*, who was supposedly the teacher of Cleopatra the Alchemist. Al-Nadim, however, attributed the authorship of the Dialogue to Maria the Jewess, leading to the supposition that these two women may have already been confused by the tenth century. See Okasha El-Daly, *Egyptology: The Missing Millennium, Ancient Egypt in Medieval Arabic Writings*, (London: Routledge, 2016), 133-134. See also, Stanton J. Linden, *The Alchemy Reader: From Hermes Trismegistus to Isaac Newton*, (Cambridge, UK: Cambridge University Press, 2003), 44-45; and F. Sherwood Taylor, *Alchemists, Founders of Modern Chemistry*, Kessinger's Legacy Reprints, (Kessinger Publishing, 2010), 57-59.
19. An Egyptian alchemist living around the time of 300 C.E., who was referred to in the letters between the alchemist Zosimos of Panopolis and his sister Theosebeia.
20. Stanton J. Linden, *The Alchemy Reader, from Hermes Trismegistus to Isaac Newton,* (Cambridge, UK: Cambridge University Press, 2003), 44.
21. F. Sherwood Taylor (1937), "The Visions of Zosimos," *Ambix*, 1:1, 88-92.
22. Lawrence M. Principe, *The Secrets of Alchemy*, (Chicago: University of Chicago Press, 2013), 16.
23. Gail Marlow Taylor, *The Alchemy of Al-Razi, A Translation of the "Book of Secrets,"* (North Charleston, South Carolina: Createspace Independent Publishing Platform, 2014), 116-119.
24. Ibid.
25. The term "water" or "waters" was used for any water-like liquid. The term "acids" was not coined until much later.
26. Gail Marlow Taylor, 116n55, 120-121.
27. Picture from R. Werner Soukup, "Crucibles, Cupels, Cucurbits," *Chymists and Chymistry: Studies in the History of Alchemy and Early Modern Chemistry*, edited by Lawrence M. Principe, (Sagamore Beach, CA: Watson Publishing International, LLC, 2007), 171.
28. Gail Marlow Taylor, 119-121. See also H. E. Stapleton, R. F. Azo, M.

Hidāyat Husain, & G. L. Lewis (1962), "Two Alchemical Treatises Attributed to Avicenna," *Ambix*, 10: 2, 41-82.

29. Gail Marlow Taylor, 119-121.
30. Gail Marlow Taylor, 127, and Stapleton.
31. Wheeler, 17.
32. H. E. Stapleton, R. F. Azo, M. Hidāyat Husain, & G. L. Lewis (1962), "Two Alchemical Treatises Attributed to Avicenna," *Ambix*, 10:2, 41-82.
33. E.J. Holmyard and Richard Russell, *The Works of Geber, Kessinger's Legacy Reprints,* (Whitefish, MT: Kessinger Publishing, 2010), 96.
34. Ibid, 98.
35. Ibid, 99-100.
36. Holmyard and Russell, 230-235.
37. Vannoccio Biringuccio, *Pirotechnia*, (Cambridge, Massachusetts: MIT Press, 1966), 183-184, 341-343.
38. Ibid, 183-184.
39. Ibid, 344.
40. Lawrence M. Principe, *The Secrets of Alchemy,* (Chicago: University of Chicago Press, 2013), 149-151. See also, Basil Valentine, Theodore Kirkringus, and Adam Goldsmith, *The Alchemy Collection: The Works of Basil Valentine, including The Triumphant* Chariot *of Antimony, The Twelve Keys, and Of Natural & Supernatural Things, Third Edition,* (Vitriol Publishing & Adam Goldsmith/L.C. Kennedy, 2013).
41. Biringuccio, 349.
42. Wheeler, 31.
43. A unit of length used in Italy. The actual length was defined by each Italian state and therefore varied.
44. Biringuccio, 348.
45. Exhalation referred to the gaseous by-products of reactions. According to Aristotle, there were two exhalations that could occur when metals or minerals formed, one was vaporous and the other smoky. The vaporous exhalation formed when sunlight fell upon water, and was moist and cold. The smoky exhalation formed when sunlight fell on dry land, and was hot and dry. (See E.J. Holmyard, *Alchemy,* 24.) According to Agricola, the transmutations of the elements in the earth produce two "exhalations," a fiery one (gasses) and a moist one (steam). (See Georgius Agricola, *De Re Metallica.*)

46. Etymoline.com, "Retort," https://www.etymonline.com/word/retort. Retrieved 15 Aug. 2023. From the French "retorte," derived from Medieval Latin "retorta," (meaning "a vessel with a bent neck," literally "a thing bent or twisted").
47. Biringuccio, 349-351.
48. John French, *The Art of Distillation, The R.A.M.S. Library of Alchemy, Vol. 15,* (Stuarts Draft, VA: R.A.M.S. Publishing, 2016), 30.
49. Ibid., 34.
50. Ibid., 34-35.
51. Ibid., 58-59.
52. Anderson, 26.
53. French, 82.
54. Ibid.
55. Ibid., 75.
56. Ibid., 84.
57. Anderson, 24.
58. French, 63.
59. Ibid., 64.
60. Ibid., 65.
61. Ibid., 65-66.
62. Ibid., 67.

Chapter 14

Hessian and Bavarian Crucibles

Originally Published in the *Hexagon*, Winter 2023, Vol. 114, No. 4

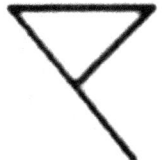

Symbol for Crucible[1]

A marriage of mining and alchemy in which crucibles played an important role, took place during the late Middle Ages and Renaissance (13th through the 17th centuries). Much of the work of the alchemists was performed in crucibles—ceramic vessels used for dry, high-temperature operations[2]. In the alchemical and early modern chymical laboratory, the crucible was the basic piece of equipment for mixing, melting, calcining, reducing, evaporating, or otherwise processing substances under high temperatures.

For the alchemists of this period, the crucible of choice was the Hessian crucible; so named for its place of origin—the Hesse region of Germany. During this period, the demand for better and more sophisticated crucibles increased because of the development of fire assay technology[3]—driven by increased mining operations and spagyric chemistry, which involved the use of high temperatures and strong reagents to refine or decompose matter. According to historical and archaeological studies, the Hessian crucibles had a wide distribution—found in various workshops in Germany, in the later medieval mints of Trondheim (Norway) and Porto (Portugal), in the Old Ashmolean[4] laboratory of Oxford (UK), and in gold-smithing as practiced in London and in colonial Jamestown, Virginia[5]. Also, the toughness—due to the presence of coarse quartz grains—of these crucibles favored the extent of their distribution. This may seem counterintuitive since the coarse grains would have induced cracks and pores in the vessels. Studies of coarse-grained ceramics have demonstrated, however, that the network of micro-cracks and inclusions hindered and dissipated fracture lines caused by mechanical or thermal stresses, preventing the fractures from propagating catastrophically. The high toughness of these crucibles would then be an asset for handling and transportation[6].

Acicular needles of mullite (Picture from *Advanced Ceramics from Preceramic Polymers Modified at the Nano Scale: A Review*. Retrieved from Researchgate.com)

So what made the Hessian crucible so desirable? From an archaeological perspective, two characteristics can be used to examine crucibles: 1) the residues adhering to the interior walls of the crucibles, leading to inferences about the materials and reactions employed, and 2) the crucible itself, from which conclusions can be drawn regarding the manufacturing processes, origins, and technical quality.

CHAPTER 14

Triangular Hessian Crucibles (Picture taken from *Tools of the Chymist*, in Chymists and Chymistry. See Note 4 below. Original photograph by S. Hape, courtesy of H.-G. Stephan)

From these studies, it has been determined that the Hessian crucibles were made of mullite ($Al_6Si_2O_{13}$)—an aluminum silicate that was not formally discovered and named until the 20th century.[7] Of course, at the time, the alchemists and chymists did not understand that the material they produced was mullite, which was synthesized following a simple recipe. The crucible-making process started with the kaolinitic clay (a clay mineral, with the chemical composition $Al_2Si_2O_5(OH)_4$, occurring throughout the world, including Germany). The kaolinite was first levigated[8] to remove impurities, and then the refined clay was mixed with fine quartzitic sand, shaped on a potter's wheel, dried, and finally fired in a kiln to high temperatures. Based on the vitrification (the process of converting into glass or a glass-like appearance) of the ceramic, and the temperature-induced mineral transformation in the clays, the estimated firing temperatures were at least 1200° C[9] and most likely would have had to be around 1400° C since the "needle" form of mullite, found in the crucibles, appears at that temperature[10].

The refining stage would ensure that any coarse particles or minerals of unpredictable chemical or thermal behavior would be removed. The presence of quartz grains within the ceramic fabric improved the toughness and the resistance to sudden temperature changes—as experienced when removing a crucible from a hot furnace—of the vessels. The high pre-firing temperatures in the potter's kiln led to the decomposition of the clay mineral kaolinite, and the subsequent formation of mullite crystals—small needle-like crystals with a very low thermal expansion—resulting in thermal shock resistance, high creep resistance, outstanding temperature strength, and exceptional stability under aggressive chemical environments. The ceramic of the Hessian crucibles, then, was composed of a network of extremely strong interlocking needles, virtually indestructible by means of mechanical stresses, fire, or strong chemical reagents.[11]

The Hessian crucible, however, was not the only high-quality crucible available. In the 15th century, a crucible made of materials from the Bavarian region of southern Germany entered the scene. In contrast to the Hessian crucible, which had a "bright" appearance, the Bavarian crucible was "dark" due to the presence of graphite. The area around Bavaria was rich in graphite deposits which through weathering processes led to the formation of clays containing abundant graphite flakes. The presence of these graphitic structures resulted in similar technical advantages as found in the Hessian crucibles. Being one of the most stable minerals at high temperatures (Graphite does not melt, but only sublimes at 3500° C.), the graphite provided the vessel's thermal refractoriness, and because of its chemical inertness would mitigate the corrosive effects of the charge material and the forming slag. In addition, the platy shape, the flexibility, and the strength of the graphite significantly enhanced the toughness and thermal shock resistance of the crucible.[12]

Symbol for Crucible[13]

During the late Middle Ages through the Renaissance, a significant increase in fire assay technology was necessitated by two factors: the need for an accurate determination of the economic viability of mining operations, and the experimentation with metals, fluxes, amalgams, reagents, and their reactions. Since mining involved a significant financial investment, a precise and accurate analysis of ores was absolutely necessary. For the alchemists, on the other hand, assaying was important to prove that a change in the amount of a substance

CHAPTER 14

was achieved as a result of a transmutational experiment. Thus, both the assayers' dry methods for determining amounts of metals in ores and the alchemists' methods of calcination and use of acids were called upon. These thermal and chemical reactions would not only attack the samples processed but also corrode the vessels as well. Only crucibles with a high-temperature resistance and a high degree of refractoriness, such as the Hessian and Bavarian crucibles, would fit the need and thus played a vital role in the union of mining and alchemy[14].

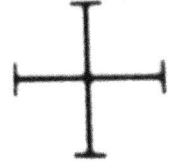

Symbol for Crucible[15]

Notes
1. Philip Wheeler, *Alchemical Symbols, Fourth Edition, R.A.M.S. Library of Alchemy Vol. 21,* (Kansas City, MO: R.A.M.S. Publishing Co., 2018), 34.
2. Crucibles can be made from any material that withstands temperatures high enough to melt or otherwise alter its contents.
3. Fire assaying is the quantitative determination in which a metal or metals are separated from impurities by fusion processes and weighed to determine the amount present in the original sample. Fire Assay Information, mine-engineer.com.
4. Elias Ashmole (1617-1692) was the founder of the Ashmolean Museum in Oxford. His most known contribution to the study of alchemy was *Theatrum Chemicum Britannicum.*
5. Marcos Martinon-Torres, "Tools of the Chymist," in *Chymists and Chymistry: Studies in the History of Alchemy and Early Modern Chemistry,* ed. by Lawrence M. Principe, Watson Publishing International: Sagamore Beach, 149-163.
6. Marcos Martinon-Torres, Ian Freestone, Alice Hunt, and Thilo Rehren, "Mass-Produced Mullite Crucibles in Medieval Europe: Manufacture and Material Properties," *Journal of American Ceramic Society,* 2008, 1-4.
7. A rare silicate mineral formed during contact metamorphism of clay minerals, which has two stoichiometric forms: $3Al_2O_3 2SiO_2$ or $2Al_2O_3 SiO_2$. Mullite- Wikipedia.
8. Ground or reduced to a fine powder. *Webster's Unabridged Dictionary of the English Language.* 2001. Random House, 2nd edition, 2001.
9. Marcos Martinon-Torres, Tools of the Chymist.
10. At around 1400° C, mullite undergoes a structural transformation from a platelet-like formation to a needle-shaped structure. See also M. Bellotto, A. Gualtieri, G. Artioli, et al., "Kinetic study of the kaolinite-mullite reaction sequence. Part I: kaolinite dehydroxylation," *Physics and Chemistry of Minerals,* Spring 1995, 22:207–214.
11. Marcos Martinon-Torres, "Tools of the Chymist."
12. Marcos Martinon-Torres, 159-160.
13. Wheeler.
14. R. Werner Soukup, "Crucibles, Cupels, Cucurbits: Recent Results of Research on Paracelsian Alchemy in Austria around 1600," in *Chymists*

and Chymistry: Studies in the History of Alchemy and Early Modern Chemistry, ed. by Lawrence M. Principe, Watson Publishing International: Sagamore Beach, 165-172.

15. Wheeler.

Acknowledgements

I would like to thank my friend Dennis Uhlig, without whose editing skills the quality of these essays would have been sorely lacking. Also, I am grateful to Dr. Brian P. Coppola, Arthur F. Thurnau Professor of Chemistry, Department of Chemistry, University of Michigan, for his critique of the essays, "Tin, Jupiter's Alchemical Influence," "Iron, Vulgar, Yet a Great Metal," "Copper, an Earthy Venus," and "Arsenic and Old Alchemy." Finally, I would like to thank my friends and fraternity brothers David "Mitch" Levings and DeWayne Gerber for their reviews of "Transmutation, The Great Work" and "Arsenic and Old Alchemy," respectively.

Index

Absorption, 63, 107, 124, 202n
Abufalah, 94-95, 118-120
Abulcasis, 242
Accidental change, 37
Acetic acid, 80, 90, 138, 150, 176, 236
Acetum, 124, 138
Adustion, 45-46
Agathodaemon, 171-173, 198n
Agricola, Georgius, 146-147, 154n, 186, 191-194, 204n, 206, 210, 214, 224-226, 231-234, 240-243, 253, 297
Albedo, 3, 43
Albert of Saxony, 38
Albertus Magnus, 40, 51, 64n, 138, 153n, 174, 183-186, 190-193, 218, 232n, 243, 254n, 295n
Alembic (Alembick), 2, 42, 62n, 184, 187, 199n, 242-243, 269-294, 295n
Al-Jawbari, 186, 190, 203n
Al-Jildaki, 220
Alkahest, 252n, 256-262, 263-264n
Al-Khwarizmi, 35, 61n, 137-138, 182-183, 202n
Al-Razi, 35, 61n, 90-93, 178-183, 187, 201n, 204n, 209, 217-218, 220, 232n, 237-238, 242, 252n, 276-280, 296n
Aludel (Alludel), 14, 108, 177-178, 278-280

Alum, 58-59, 87-91, 99, 112n, 169-170, 176, 181, 187, 206, 209, 212, 201-217, 219-220, 226, 232-233n, 238, 240, 243, 245, 252n
Amalgam, 48, 58, 59, 61n, 90-93, 98, 110, 156, 302
Amalgamation, 59, 90, 91, 98
Ambix (alembic), 270
Ammonia, 35, 182
Antimony, 4, 25, 31n, 74, 76, 82n, 105, 109, 128-130, 153n, 155-163, 165n, 201n, 297n
Antimony, Regulus of, 109, 156, 159, 161
Antimony, Saturn of, 76, 82n
Aphrodite, 2, 134
Apollo, 2
Aqua fortis, 76, 99-101, 108-109, 124, 220, 236, 239-241, 246, 253n, 257, 288
Aqua nitri, 239
Aqua regia, 100, 108-109, 236, 245-250, 253n, 254n
Aqua valens, 239-240
Aquarius, 2
Aquavitae, 48
Aquosity, 222
Archeus, 24, 30n, 143-144
Ares (Aries), 2, 159
Argentvive, 15, 45-47, 97, 120, 141, 143

Aristotle, 5, 12-13, 21, 24, 36-37, 168, 190-191, 197n, 204n
Arnold, 31n, 295n
Arsenic (Arsenick), 25, 35-36, 59, 62n, 84, 88, 91, 112n, 119, 121, 129, 143, 153, 160, 163, 166-204, 241, 242, 273, 276
Arsenic sulfide, 35, 88, 112n, 121, 167, 169, 170, 174, 177-181, 187, 190, 192, 194, 197n
Arsenolite, 167
Arsenopyrite, 179, 194, 196n, 202n
Ashmole, Elias, 300, 304n
Assation, 45
Atomism, 18n, 36-38, 61n
Atramentum sutorium, 208, 213, 218, 224-227, 233n
Auric chloride, 100
Auripigmentum, 168, 183, 191, 194
Avicenna, 51, 92-94, 110, 112n, 138, 232n, 280, 297n
Bacon, Roger, 31
Bain-marie, 199n, 272, 274, 282, 292, 295n,
Balneum maris, 222
Bell-Metal, 105
Biringuccio, Vannoccio, 100-104, 113-114n, 240, 253n, 283-287, 297n, 298n
Bitumen, 193, 197n
Black Crow, 2
Boerhaave, Herman, 257
Boraxes, 35, 178
Boyle, Robert, 13, 29, 82n, 113n, 157, 164n, 245, 250, 258-260, 262, 263-265n

Brass, 84, 89, 105, 135-140, 144-145, 147, 211, 227, 276, 282, 292-293
Brimstone, 14, 194
Bronze, 84-86, 111n, 135-138, 166, 196n, 214, 272
Cadmia, 87, 137, 169, 214, 225, 231n, 276
Calcination, 2, 16, 72, 91, 96-97, 106, 137, 140-141, 143-144, 146, 148, 215, 222, 256, 271, 303
Calcium carbonate, 106, 211, 231n
Calx, 16, 45, 91-92, 96, 107, 141, 143, 187, 246
Cancer, 2, 99, 147
Capricorn, 2
Caput corvi (Head of the Crow), 42
Caput mortuum, 222-223, 244, 288
Cassiterion, 78
Cement royal, 51
Cementation, 51, 56-58, 95, 140, 143, 144, 153n
Ceruse, 72, 78-79
Cerussite, 78, 87
Chalcanthite, 206-207, 214, 230-231n
Chalcitis, 135, 209-211, 214-217, 224-227, 233n
Chalcogen, 11-12, 18n
Chalcopyrite, 60, 139, 207, 227
Chalk, 106, 114n, 154n, 211
Chalybs, 129-130, 160, 163
Chaos, 26, 128-130, 159-160, 163, 259
Charcoal, 79, 109, 140, 174, 185, 215, 261, 273
Chinese iron, 35, 178, 181-182

INDEX

Cinnabar, 14, 89, 90, 170, 176, 178, 192, 194, 197n, 269

Circulation, 285, 291

Cleopatra the Alchemist, 271, 275

Cobalt, 69, 194

Cohobation, 285, 286

Colcothar, 76, 217-220, 223, 226-227, 244

Cold still, 288-289

Compositio, 12, 13

Congelation, 2, 50, 232n

Copper chloride, 100

Copper, Crocus of, 148-149

Copper, Flower of, 137, 146, 227

Copper, Scoria of, 137

Copperas, 207, 219, 228-229, 242

Corpuscle, 13-16, 38, 44, 47, 61-62n, 109, 150, 259-261

Crocus martis, 219, 223

Cronus, 70, 81n

Crucible, 44, 48, 106, 109, 148, 153n, 186, 190, 204n, 278, 299-303, 304n

Cucurbit, 107, 113n, 270-272, 276-293, 296n, 304n

Cupel, 139, 204n, 296n, 304n

Cupellation, 56, 77, 81n, 95, 140

Decoction, 124

Diana, 2

Dibikos, 272, 274-275

Digestion, 2, 27, 109

Dioscorides, 78, 128, 164, 168, 192, 209-210, 212-218, 227, 228, 231n, 269

Dissolution, 2, 21, 47, 100, 108, 109, 118, 131, 149, 156, 174, 199n, 213, 238

Distillation, 2, 27, 28, 32n, 115n, 124, 150, 172, 184, 219, 221, 222, 234n, 236, 238, 240-245, 252n, 256, 257, 269-294, 295n, 298n

Dove, 160

Dragon, 3, 111n, 129, 130, 160, 248-249, 252n

Dulcos, Samuel, 257, 263n

Eagle, 2, 248-249

Elixir, 3, 41, 43, 53, 54, 85, 90, 92-93, 121, 129, 156, 163, 175, 183, 184, 186, 242, 257

Emanationism, 3, 22

Emerald Tablet, 6, 31n, 41

Extraction, 11, 26, 55, 56, 112n, 124, 175, 180, 224, 280

Fermentation, 2, 40, 185

Ferric chloride, 100

Ferrum potabile, 125

Fixation, 2, 15, 96, 97, 120, 184, 190, 204n

Flux, 79, 139, 146, 182, 193, 273-274, 285, 291, 302

French, John, 238, 241, 244, 250, 252n, 283, 287, 292, 293, 298n

Fuller's earth, 261

Galen, 22, 24, 30n

Galena, 74, 77, 78, 83n, 105

Gemini, 2

Giles of Rome, 40

Glance, 78, 163, 209

Glauber, Johann, 239, 241, 252n, 253n

Graphite, 78, 302

Green Lion, 2, 160, 221, 222, 242

Gypsum, 35, 179, 192, 197n, 201n

Hematite, 35, 58, 179

Hermes Trismegistus, 6, 27, 31n, 41, 202n, 296n
Hermetic Tradition, 27, 71, 81n
Hollandus, Isaac, 31n
Hollandus, Johann, 6, 73, 75-76, 82n
Hydrochloric acid, 100, 228, 237-239, 241, 246, 249, 250, 252n, 254n
Hylomorphism, 36
Hylozoism, 41, 62n
Iatrochymistry (Iatrochemistry), 123, 148, 151, 256, 260
Ignis Aqua, 258
Imbibition, 45, 124, 184, 202n
Incineration, 2
Inversion, 50, 55-56
Iron, Oil of, 125
Jabir ibn Hayyan, 18n, 95, 118, 174, 186, 200n
Jupiter, 2, 48, 69, 71, 86, 93, 95, 96, 98, 100, 101, 104, 105, 107-110, 119, 137, 139, 141, 246
Jupiter Diaphoretick, 106, 109
Jupiter, Flowers of, 106-108
Jupiter, Magistry of, 106, 107
Jupiter, Salt of, 106, 107
Kaolinite, 301, 302, 304n
Kerotakis, 173, 199n, 273-276
Kohl, 35, 179, 182, 201n
Lapis lazuli, 35, 179, 182, 197n, 201n
Lavament, 46, 72, 143
Lead sulfide, 77
Lemery, Nicholas, 105-108, 115n, 116n, 130-131, 133n, 147-151, 154n, 184, 185
Leo, 2, 159, 222, 242
Libavius, Andreas, 31n, 238, 244
Libra, 2
Litharge, 78, 79, 82n, 83n, 157, 164n, 165n
Llull, Ramon, 240, 295n
Logos, 3, 23
Luting, 150, 178, 284, 287, 288, 291
Magnes, 129, 130, 160, 163
Magnesia, 35, 87, 169, 179, 192, 197n, 201n
Magnesium, 25, 108, 115n, 182, 199n
Magnetite, 87, 128, 169
Maier, Michael, 4, 62n, 160, 271
Malachite, 35, 135, 179, 196n, 201n, 207
Mappa Clavicula, 57, 58, 59, 65n
Marcasite, 35, 60, 91, 119, 144, 179, 182, 201n, 214, 227, 230n
Maria the Jewess, 34, 171, 198n, 199n, 271, 275, 296n
Mars, 2, 16, 48, 69, 73, 97, 98, 100, 109, 117-131, 132n, 135, 137, 141, 142, 145, 157-159, 219, 242, 243
Mars, Crocus of, 123-125, 243
Mars, Saffron of, 130
Mars, Vitriol of, 130-131
Massicot, 79, 83n
Melanteria, 209-211, 227
Mercury-Sulphur Theory, 18n, 22, 95
Minima, 13, 36, 38-41, 62n
Minium, 58, 75, 76, 79
Misy, 135, 209-215, 219, 224-227, 231n, 233n
Mixis, 12, 13
Monism, 259
Moon, 2, 69, 98, 100, 101, 119, 129, 135, 137, 198n, 243

INDEX

Mullite, 300-302, 304n
Multiplication, 2, 48, 164n
Mundification, 46
Natron, 35, 57, 58, 59, 168, 172, 173, 181, 182
Natural sympathy, 88
Neoplatonism, 22
Newton, Isaac, 13, 115n, 159
Nicander, 78, 168, 197n
Nigredo, 3, 42
Nitre, 25, 99, 106, 108, 149, 157, 190, 220, 241, 250
Nitric acid, 100, 115n, 124, 149, 150, 219, 238-250, 253n, 254n, 257, 261
Oldenburg, Henry, 257
Oleum vitrioli, 223, 244
Olibanum, 176
Orpiment, 59, 87, 167-187, 191-194, 196n, 197n, 199n
Ovum philosophicum, 42
Peacock, 3, 43
Pelican, 2, 285, 291
Pewter, 84, 289
Philalethes, 4, 62n, 128-130, 133n, 156, 158-160, 165n, 258, 264n
Philosopher's clay, 42
Philosopher's Stone, 42, 121
Philosophical Gold, 156, 163
Pirotechnia, 100, 102, 104, 113n, 114n, 240, 283-286, 297n
Pisces, 2
Pliny the Elder, 85, 135, 152n, 170, 198n, 209, 231n, 232n
Plumbago, 76, 78, 79
Plumbum, 71, 77, 78, 81n
Plusquamperfection, 50

Potash, 185, 203n, 217, 260, 261
Potassium carbonate, 107, 108, 184, 185, 250, 260, 261, 262, 264n
Potassium nitrate, 25, 108, 189, 190, 211, 241, 244, 245, 249
Prime matter (primeval matter), 3, 22, 23, 30n, 32n, 40, 51, 52, 88
Projection, 2, 44, 46, 48, 54, 63n, 189, 242
Pseudo-Democritus, 19n, 36, 37, 55, 61n, 64n, 87-88, 111n, 157, 164n, 169-170, 198n, 209, 212, 231n
Pseudo-Geber (Geber), 17n, 61n, 72-73, 76, 95-97, 101, 119-120, 121, 139-142, 151, 153n, 186-187, 219, 232n, 239, 240, 242, 243, 246, 254n, 281, 282
Pseudo-Khalid ibn Yazid, 89
Pseudo-Maimonides, 98
Purgation, 4, 50, 51, 143, 160
Putrefaction, 28, 221, 223
Pyrite, 60, 74, 77, 78, 83n, 87, 91, 93, 105, 135, 139, 144, 153n, 157, 169, 179, 197n, 206, 214, 219, 224, 226, 227, 230n, 231n, 233n
Pyrotartaric acid, 260, 263n
Quicksilver, 14, 35, 45, 46, 52, 59, 63, 90, 94-98, 103, 105, 118, 119, 128, 137, 156, 160, 182, 237, 262
Quintessence, 127, 256, 263n
Realgar, 87, 88, 167-179, 183, 185, 187, 191-194, 196n, 197n, 199n
Redox reaction, 174
Refuse, 78, 128
Retort, 25, 43, 150, 199n, 221, 236, 238, 239, 241, 244, 245, 249, 250, 269, 287, 288, 298n

Ripley, George, 128, 133n, 156, 158, 165n
Rosarium Philosophorum, 53, 64n
Rosenhut, 289
Rubedo, 3
Rulandus, Martinus, 6, 65n, 71-78, 81n, 82n, 83n, 105, 114n, 128, 133n, 143, 147, 154n, 194, 205n, 206, 226-228, 233n, 234n, 252n
Sagittarius, 2, 159
Sal Ammoniac, 35, 79, 99, 100, 108, 124, 161, 178, 181, 183, 199n, 219, 237, 240, 242, 246, 249, 252n, 253n, 261, 277
Salt, 20-28, 31n, 35, 45, 57-59, 75, 87, 90, 91, 99, 106-109, 115n, 121-131, 143, 144, 145, 150, 159, 163, 165n, 178, 180, 181, 182, 185, 187, 192, 209, 211, 231, 237-239, 241, 246, 247, 250, 258-262, 278
Salt, Spirit of, 108, 109, 236-241
Saltpeter (Saltpetre), 211, 219, 225, 240, 242, 243, 244, 249, 261, 262
Sandarach, 170-174
Saturnia, 129, 158, 159, 163
Scaliger, Julius, 39, 255n
Scoria, 78, 128, 137, 162
Scorpio, 2
Sedacer, Guillaume, 187-188,
Semina, 36, 39-41, 50, 55, 62n, 259, 260, 264n
Sendivogius, Michael, 250
Sennert, Daniel, 39, 250
Separation, 2, 28, 31n, 32n, 99, 124, 144, 149, 189, 223, 236, 244, 251, 256, 262, 281, 287
Sideritis, 128

Silver nitrate, 100, 102
Sodium Chloride, 25
Solen, 270, 283
Sory, 135, 209-218, 224-227, 233n
Spagyria, 28, 32n
Spiritus salis, 237
Spirits of Saturn, 79
Spiritus vitrioli, 223
Stannic chloride, 100, 102
Stannite, 97
Starkey, George, 62n, 76, 82n, 113n, 128, 130, 133n, 156-157, 160, 164n, 257, 258, 261, 262
Stellate regulus, 76
Stibnite, 4, 74, 156-163, 165n
Sublimation, 2, 14, 19n, 28, 44-47, 91, 93, 106, 107, 121, 131, 142-144, 167, 169, 174-177, 180, 181, 187, 193, 199n, 237, 252n, 256, 257, 269, 271, 275-279, 283
Substantial change, 37, 55, 58
Sulfate, 25, 37, 60, 121, 124, 139, 144, 169, 176, 187, 210n, 206-228, 230n, 231n, 232n, 234n, 237, 240, 242-244, 254n, 261
Sulfuric acid, 100, 131, 170, 222, 223, 228, 232n, 238-245, 250, 254n
Sulphur, 14, 15, 16, 18n, 20-25, 28, 31n, 49, 58, 71, 75, 90-97, 105-109, 120, 121, 127, 129, 130, 140-143, 148, 156, 162, 169, 173, 182, 184, 187, 188, 193, 212, 249, 272, 280
Sulphureity, 15, 16, 45, 96, 141
Summa Perfectionis, 13, 38, 47, 61n, 62n, 63n, 64n, 72, 81n, 95, 142, 153n, 186, 203n, 220, 232n, 239, 242, 254n, 281

INDEX

Sun (Sol), 2, 69, 87, 99-102, 119, 129, 135, 146, 159, 213, 224
Tachenius, Otto, 32n
Talc, 35, 45, 153n, 179, 180, 201
Tartaric acid, 131, 185
Tarter, Oil of, 107, 108, 184, 185, 190
Taurus, 2
Teallite, 97
Terrestreity, 45
Theophrastus (Aristotle's pupil), 168, 197n
Thölde, Johann, 221, 244, 254n
Tin, 84-116
Tin, Butter of, 101
Tin, Common, 105
Tin, Plate, 105
Tincture, 2, 31n, 41, 48, 51, 53, 54, 58, 89, 124-127, 130, 131, 143, 145, 173, 184, 189, 190, 221, 222, 280, 281
Transmigration, 36
Transmutation, 14, 33, 36, 37, 40, 41, 43, 44, 47-60, 63n, 64n, 72, 73, 80, 82n, 84, 88, 90, 92, 110, 111n, 119, 120, 122, 123, 131, 154n, 156, 158, 167, 181, 183, 188-195, 241, 256, 259, 260, 262, 264n, 276, 297n, 303
Tree of Saturn, 80
Tria Prima, 20-32, 118, 257-259
Tribikos, 199n, 272-274
Trinitarianism, 22
Turquoise, 35, 179, 182
Tutia, 35, 121, 139-143, 179, 201n
Twinning, 84, 85, 94, 110
Universal solvent, 257, 263n, 264n

Valentine, Basil, 3, 31n, 73, 74, 82n, 125-127, 157, 158, 163, 165n, 221, 222, 234n, 244, 247, 249, 285, 297n
Van Helmont, Joan Baptiste, 28, 40, 62n, 258-262, 263n, 264n
Venus, 2, 16, 48, 60, 69, 73, 97, 100, 101, 119, 127, 134-151, 189, 221, 223
Verdigris, 138, 144-147, 150, 227, 242
Vinegar, 46, 47, 59, 72, 79, 90, 91, 107, 123, 124, 138, 145, 147, 150, 163, 165n, 170, 175-181, 187, 231n, 236, 261
Virgin's wax, 50
Virgo, 2
Vitriol, Roman, 207, 243
Vitriol, Arcanum of, 222
Vitriol, blue, 206, 207, 214, 228, 243
Vitriol, Cyprus, 219, 240
Vitriol, green, 35, 207, 208, 217, 220, 227, 228, 240
Vitriol, Oil of, 131, 222, 228, 236, 239, 241, 242, 245
Vitriol, white, 35, 208, 217, 220, 225, 228, 234n
Von Suchten, Alexander, 156, 164n
Wine, Spirit of, 115n, 127, 131, 261, 264n, 291
Wolf, 73, 157-163
Zinc, 60, 80, 135, 139, 146, 179, 182, 196n, 199n, 201n, 206-208, 214, 225, 231n, 234n, 276
Zodiac, 118
Zosimos, 6, 34, 90, 171, 173, 174, 198n, 199n, 271-276, 296n

About the Author

Keith Schuette is a 1977 graduate of the University of Missouri-Rolla (currently the Missouri University of Science and Technology), where he earned a BS degree in Metallurgical Engineering. He also holds a Master of Divinity degree from Eden Theological Seminary in St. Louis. Now retired, his career involved employment as a metallurgist, a quality manager, and a part-time instructor. A father of a daughter and a son, and a grandfather of six, he currently resides with his wife, Linda, in West Bend, Wisconsin. He is also the author of *Beta Delta of Alpha Chi Sigma (A History)*.

www.ingramcontent.com/pod-product-compliance
Lightning Source LLC
Chambersburg PA
CBHW071735150426
43191CB00010B/1584